Teach Your Children Tables

Third Edition

BILL HANDLEY

Wrightbooks
A Wiley Brand

First published in 2015 by Wrightbooks
an imprint of John Wiley & Sons Australia, Ltd
42 McDougall St, Milton Qld 4064

Office also in Melbourne

Typeset in 11/13 pt Trump Mediaeval LT Std

© Learning Unlimited Australia Pty Ltd 2015

The moral rights of the author have been asserted

National Library of Australia Cataloguing-in-Publication data:

Author:	Handley, Bill
Title:	Teach Your Children Tables / Bill Handley
Edition:	3rd Edition
ISBN:	9780730319634 (pbk.)
	9780730319641 (ebook)
Subjects:	Multiplication—tables—study and teaching (primary)
	Arithmetic—study and teaching (primary)
	Numeracy—study and teaching (primary)
	Mathematics—study and teaching (primary)
Dewey Number:	372.72

Cover Design: Wiley

Cover Image: © iStock.com/gbh007

Printed in Australia by Ligare Book Printer

10 9 8 7 6 5 4 3 2 1

Disclaimer
The material in this publication is of the nature of general comment only, and does not represent professional advice. It is not intended to provide specific guidance for particular circumstances and it should not be relied on as the basis for any decision to take action or not take action on any matter which it covers. Readers should obtain professional advice where appropriate, before making any such decision. To the maximum extent permitted by law, the author and publisher disclaim all responsibility and liability to any person, arising directly or indirectly from any person taking or not taking action based on the information in this publication.

Also by Bill Handley

Speed Mathematics:
Secrets of Lightning Mental Calculation, 3rd edition

Speed Maths for Kids:
Helping Children Achieve Their Full Potential

Fast Easy Way to Learn a Language

Speed Learning for Kids

Contents

Preface

The previous editions of *Teach Your Children Tables* have enjoyed acceptance around the world. I have received mail and email from parents, educators and children telling me how much the book has helped them. When I have written follow-up books, I have noticed that my explanations have undergone subtle changes. In particular, I've made a number of changes to the way I teach very young children. I have modified my explanations and included other methods and explanations to meet the difficulties of explaining multiplication and addition to children with no understanding of the concepts.

In this third edition of *Teach Your Children Tables* I have included new chapters on factors, squaring numbers, learning beyond the tables, an appendix on why the methods work and, one of my favourites, the encyclopaedia salesman puzzle. This puzzle is a great example of how to use my suggestions for solving problems, or puzzles, when you have no idea how to begin. Solving these puzzles is the most effective and enjoyable way I know to develop an understanding of the properties of numbers and to learn problem-solving skills.

I have also included a chapter that looks at an easy method of direct multiplication. I have not seen this exact method taught anywhere else.

Finally, I have provided advice for parents and teachers on the best ways to teach these methods.

Bill Handley
Melbourne
December 2014
bhandley@speedmathematics.com

www.speedmathematics.com

Acknowledgements

I owe so much to so many for the existence of this book.

First, Geoff Wright, founder of Wrightbooks, who phoned me after hearing me in a radio interview and said, 'Write a book and I will publish it'. Geoff invited me to spend a day speaking with the staff of Wrightbooks and, at the end of the day, he said, 'Write a book covering what you told us today'. This was the book. He was prepared to take a risk and it paid off. Thank you, Geoff.

Thank you also to Geoff's daughter, Lesley Beaumont, who gave me heaps of guidance, advice and encouragement with my earlier books. Michael Hanrahan, who edited the first and second editions of this book and provided the computer-generated illustrations.

I would like to thank the people at Wiley directly connected with the third edition of this book: Lucy Raymond, Meryl Potter and Danielle Karvess. They have all played an important part in the preparation of this book. There are so many others at Wiley who deserve thanks. I am grateful for your support.

Thank you to the educators around the world and to the lecturers at teachers' college who have encouraged me to develop and refine my methods. I also want to thank and acknowledge my students, whom I have taught over the years and who have forced me to modify and improve my methods.

I would like to dedicate this book to my grandchildren and to students everywhere.

Introduction

Imagine your children going to school tomorrow, confident in their ability to do maths and get the right answers. Imagine your children being able to check their answers and correct any mistakes before the teacher, or anyone else, sees them. Imagine your children having a reputation for getting 100 per cent for most maths tests. What would that do to your children's self-esteem? What would it do for their confidence? Wouldn't your children enjoy school more? You can teach your children to master the basic multiplication tables in around 15 minutes. They don't need a super mathematical brain to do this: they just need an easy method, which I supply in this book.

People who excel at mathematics use better strategies than the rest of us—they don't necessarily have better brains. This book teaches simple strategies that can have you and your children multiplying large numbers in your heads, checking your answers and enjoying mathematics in no time at all.

And here is a secret. People equate intelligence with mathematical ability. In other words, if you are able to do lightning-fast calculations in your head, people will think you are intelligent in other areas as well. Students who learn these methods often improve markedly in other subject areas and find they enjoy school much more. And mathematics usually becomes the child's favourite subject.

Speed and accuracy

One of my most important rules of mathematics is this:

The easier the method you use to solve a problem, the faster you will solve it with less chance of making a mistake.

It is an unfair rule, but it is a rule just the same. Why is it unfair? Imagine your teacher asks the class to solve 56 minus 9 in their heads. Some students try to solve the problem the same way they would with pencil and paper. They say to themselves:

'Six minus nine won't go so we borrow from the tens column. That means we now subtract nine from sixteen...'. This method is more complicated than simply subtracting 10 (one too many) and then adding back the 1 to get the answer. Or subtracting 6 for an answer of 50, and then subtracting another 3.

The more complicated the method you use, the longer you take to solve a problem and the greater the chance of making a mistake. The people who use better methods are faster at getting the answer and make fewer mistakes, while those who use poor, complicated methods are slower at getting the answer and make more mistakes. It doesn't have much to do with intelligence or having a 'mathematical brain'.

Believe in yourself

What is the result? The children with the 'better' methods believe they are more intelligent than the other students, and the teachers and other students believe it as well. It has been shown that this belief can have a positive effect on a child's learning. In several countries, tests were conducted with school children. At the beginning of a school year, teachers were told that a number of children had exceptionally high IQs and were given a list of their names. At the end of the year, it was found that those children had outperformed their classmates. The names, however, had been chosen at random. The test was conducted to see how much a teacher's expectations affected a child's performance.

In Israel, it was observed that many children of families arriving to work in Kibbutzim tested poorly in IQ tests on their arrival. Twelve months later, those same children tested significantly higher. A number of reasons were put forward for their improvement. The children were provided with encouragement and support, creating a positive learning environment. They were also provided with opportunities to develop their intelligence that they had not experienced before.

Make learning fun

The methods I teach are not only fun to use, they are also easy to learn. I walk into a class of primary school children and teach them the basic multiplication tables in half an hour—up to the 20 times tables. They are then multiplying numbers like 95 times 95 in their heads faster than you can punch the numbers into a calculator. And they love it. If there is an easy method to teach the multiplication tables, and basic mathematics and number facts, how can we justify teaching a more difficult method and withholding the easy method from the students?

This book provides more than just techniques for fast calculation. Students who learn the methods are more inclined to think for themselves rather than follow a set formula. They don't ask themselves, 'How was I taught to solve this problem?' but rather, 'What is the easiest way to solve this problem?'

The methods develop strategies for general problem-solving and encourage lateral thinking skills. If you don't know or haven't been taught how to solve a problem, you will work out your own method. Sometimes those methods are clumsy or complicated and difficult to use in practical situations. We often don't realise that there is a much easier method for performing the calculation. Many students have already learnt the methods I teach in this book and have developed reputations for being mathematical prodigies. They have excelled, not only in mathematics, but in other subjects as well.

Tip

This is a fun book. It is not intended to be too technical.

In this book, you and your children will learn:

- to master the multiplication tables up to, and greater than, the 20 times tables, with no rote memorisation

- to check your answers so you can develop a reputation for never making a mistake

- easy strategies for addition and subtraction

- the properties and characteristics of numbers.

While you are learning and using the strategies, we will be working with factors (explained in chapter 8), and positive and negative numbers. We will also look at some fun short cuts you can play with, and some puzzles and ideas for problem-solving. We will lay a foundation for the future and have fun with mathematics along the way.

I don't want anyone to regard this book as hard work, but rather a book to have fun with and enjoy.

Now read on and enjoy yourself as you make mathematics your favourite subject—and your children's favourite subject. By the time you have finished this book, you and your children will be able to complete the practice sheets at the back of the book with ease.

How to use this book

If you are a parent or a teacher, you can enjoy extraordinary success teaching children the basics of mathematics. It has been my experience that children who learn their multiplication tables using this method also master their basic number facts and mathematical concepts much more quickly than students taught by traditional methods. And children work harder because they see results for their efforts. Here are my suggestions for effective lessons.

Use the practice sheets

Photocopy the practice sheets in appendix D or print them from my website at www.speedmathematics.com. (From the website, go to the page of resources for teachers and homeschoolers. You will need Acrobat Reader to be able to see and print the pages, which can be downloaded for free from the internet.) Do the exercises with your child. Tell your child how much you enjoy playing with the methods.

Take your child gradually through the book. There is no need to push the pace or to put the child under pressure. Give plenty of encouragement. Make a game of it—make it an adventure. You are learning with your child. Tell your child that you are excited about what *you* are learning.

When the child makes mistakes

If your child makes a mistake, don't make a big deal about it. Tell your child, 'Of course you will make mistakes. This is all new stuff. Everyone makes mistakes in the beginning'. I tell my students, make your mistakes before you begin to show off to your friends. Everyone makes mistakes when they start, and it's best to make them all now so you don't embarrass yourself in front of your friends. Just try again, and practise until you get the right answer every time.

At what age can children begin learning these methods?

Many parents are afraid to teach their children for fear they might be 'pushing' them. If a child wants to learn, you are definitely not 'pushing' him. You will just frustrate the child if you refuse to teach or answer his questions. I have found that children as young as four or five want to learn and are able to understand the concepts of addition and multiplication.

No-one is too old to learn the methods in this book. Many parents have told me that they were more excited than their children to learn the strategies. Teachers and student teachers are usually my most enthusiastic students.

Will it cause trouble at school?

This is really a double question. The first question is, if the child learns mathematics before starting school, will the teachers be upset? The short answer is no. Children begin school with a wide range of abilities. The better prepared a child is, the easier he will be able to cope with schoolwork.

The second question is, if my child is already at school, won't it confuse him to be taught a different method for solving problems? Many parents share this concern. However, these methods complement what is taught in the classroom. The school requires that students learn the multiplication tables. It doesn't specify *how* the students do so. If a child is asked to subtract 9 from 64, the teacher is interested in the child giving the correct answer. The teacher doesn't know if the child subtracts 10 and then adds 1 to the answer, or if the child subtracts 4 and then another 5 to make 9. Teachers will understand that children have different strategies for finding the answer.

I have found that children who use the methods to learn their tables in this book are not confused or disadvantaged because they used a different method to learn them. Most of the methods taught in this book are invisible. Teachers are unlikely to be concerned if the child chooses not to divulge the method he uses. You can even encourage your child to be mysterious about how he is able to perform the calculations. The rest of the class will simply believe your child is a genius. This is exciting stuff for any child.

Key points

- Work your way through the book together with your child.

- Treat mathematics as play rather than work.

- Show enthusiasm for what you are learning.

- Let your child know you are excited and looking forward to the next session together.

Part I
Learning the method

Part I describes a simple method for teaching multiplication and enables students to master their tables in less than an hour by calculating the answers.

This is a simple method I have developed for teaching multiplication and basic number facts. It is the best way I know for children and adults to achieve instant success, which gives the student a high level of motivation.

Chapter 1
Teach your child tables

What is multiplication? How can you teach your child what the word means? If you are going to teach your child the multiplication tables and how to multiply, it is necessary to first teach her what multiplication is.

This is a method that I have found useful. I begin by holding up one hand and asking the child how many hands. Then I hold up both hands and ask how many. Then I hold up three fingers on one hand and ask how many fingers.

Next I hold up three fingers on each hand and ask her how many fingers. I tell her she can count the fingers if she likes. If she counts correctly, she will get the correct answer of six. I explain that she can add the fingers on each hand—that three fingers plus three fingers makes six fingers. That is addition.

I tell her she can also say two hands with three fingers, or two lots of three fingers make six fingers. That is multiplication. So three plus three and two threes are both the same—they both equal six.

You could ask someone else to hold up three fingers on each hand, too, so there are four hands altogether. Then you could either say three plus three plus three plus three or, simply, four threes. The child could also count the fingers to find that four threes make twelve fingers.

It's useful to point out that the more hands and fingers you have to count, the greater the chance of making a mistake. For example, when you have to count fingers on eight or nine hands it becomes hard work and you have even more chance of making a mistake.

You can use other examples. How many marbles in a glass? How many marbles do you have if you have three glasses with four marbles in each glass? This is all multiplication.

Multiplication is simply a short, easy way to write and do addition. It is addition written in shorthand. For instance, 3 times 7 means 7 plus 7 plus 7, or three sevens added together. Seven times 8 means 8 plus 8 plus 8 plus 8 plus 8 plus 8 plus 8, or seven eights added together. Two times 6 means two sixes added together, or 6 plus 6. A child needs to be able to understand this concept to use the methods taught in this book. That doesn't mean she must know the answers to 6 plus 6 or 4 plus 4.

I was asked to teach maths to a 14-year-old boy and, at our first meeting, I asked him if he knew his tables.

He said, 'Yes, up to the two times table'.

I said, 'What are two eights?'

He shook his head and said, 'I don't know them that well'.

Within a week, he was multiplying 96 by 96 in his head. Because the methods were working for him, he practised them and so he learnt his two and three times tables. He asked me if I could teach him to beat his friends, who do their multiplication with a calculator.

So long as the child knows what two sixes means, and how to calculate two sixes, he will be able to make good use of the methods taught in this book. A five-year-old boy who hadn't yet started grade one mastered multiplication using these methods. When he had to multiply two fours, he would extend four fingers on each hand and count the fingers. It wasn't long before he knew two threes and two fours. But he was multiplying two fours to calculate $96 \times 98 = 9408$. Since learning this, he has been regarded as a maths prodigy at his school.

Using my methods, a child can memorise the tables up to around the 13 or 14 times tables in a couple of months through repetition. Not repetition in reciting the tables, which is boring, but repetition in calculating the answer, which is fun. It is fun because the child knows that no-one else in the class, or maybe even in the school, can do the calculation so easily. Certainly, the child wouldn't expect that other children could multiply 96 times 96 and give an immediate answer seemingly off the top of their head.

We use maths every day

Some children have said to me, 'When do we need this? When do we ever have to multiply or divide in real life?'

It is a fair question and should be answered. In fact, it must be answered if you want to keep the child motivated. Following are some situations you can present to children where basic maths skills are needed:

- *Situation one*—If three children get $5 pocket money each week, how much do they get all together?

  To solve this easily, they need to be able to multiply. This might be important if they want to pool their money to buy something. The answer is $15.

- *Situation two*—If you want to buy 16 DVDs at $14 each, how much money will you need to buy them?

  If they learn the simple method to multiply numbers in the teens, taught in these first few chapters, they will be able to give an immediate answer of $224.

- *Situation three*—If we are driving towards a town 240 km from home and we have driven 60 km, how far do we still have to drive? The answer is 180 km.

- *Situation four*—If your child earns $5 per week in pocket money, how long will it take to save $40? If the child spends $1 a week for essentials, how long will it take?

 Tip

Encourage children to use the methods when they are shopping, cooking or making models. Show them it is difficult to go through a whole day without using some maths.

Working with your child

The best way to teach your child basic maths and the tables is to do the exercises together. If you get excited about what you are doing, this will in turn excite the child. Ask her how many children in her class could do the problem so quickly in their head. Ask if the teacher could do the problem.

Don't make firm rules. Don't say, 'You are going to spend 15 minutes each night after school doing these tables'. This makes it sound like work for the child. On the other hand, you can say, 'Let's do this together each night for about 15 minutes. I want to learn this. If you are willing, so am I'.

In one case, a father told me he couldn't get his daughter to cooperate. I told him to learn the methods himself and then show his daughter what he had learnt. He did so. She asked him, 'How did you do that? Teach me'. He said, 'No, you will have to wait until I have finished'. She pleaded with him until he relented. She was desperate to begin.

The methods are fun because the results seem unbelievable. If you suspect your child might be reluctant to work with the methods, don't offer the book, simply ask, 'How would you like to master your tables in the next 10 minutes?' Or you could say, 'I learnt a great way to learn the multiplication tables today. You can master your tables in five minutes. Let me show you this great trick'. That is a good, positive approach. (Actually, I dislike people calling these methods a 'trick'. To me, a trick implies deception; you aren't really doing what you appear to be doing. In this case, you are doing exactly what you appear to be doing.)

Spend fifteen minutes a day

When the child is working with the methods and seeing results, you can then say, 'Let's spend 15 minutes a day together, three or four days a week, working on your tables.' Get the child's cooperation and give plenty of encouragement. These methods are very easy to learn and to use.

As you learn with your child, you will be motivated to play and experiment with the strategies as well. This will also help to keep your child motivated.

Now, turn to the next chapter and become a mathematical genius.

Key points

- Multiplication is a short, easy way to write and do addition.

- We use maths every day.

- Parents should learn these methods with their children.

Chapter 2
Multiplication — getting started

How would you like to master your tables, up to the 10 times table, in around 15 minutes? And your tables up to the 20 times table in less than half an hour? Using the methods I explain in this book, it is possible to master the multiplication tables, up to the 10 times table, in one lesson. I only assume you know the two times table reasonably well, and that you can add and subtract simple numbers.

Multiplying numbers up to 10

We will begin by multiplying numbers up to 10 times 10. Let's take a look at an example: 7×8.

Write '$7 \times 8 =$' on a piece of paper and draw a circle below each number to be multiplied, as shown here.

$7 \times 8 =$
○ ○

Now go to the first number to be multiplied, 7. How many more do you need to make 10? The answer is 3. Write 3 in the circle below the 7. Now go to the 8. What

9

do we write in the circle below the 8? How many more to make 10? The answer is 2. Write 2 in the circle below the 8. Your work should look like this:

$7 \times 8 =$
③   ②

Now subtract diagonally. Take either one of the circled numbers, 3 or 2, away from the number, not directly above, but diagonally above, or crossways. In other words, you either take 3 from 8 or 2 from 7. Either way, the answer is the same, 5. This is the first digit of your answer. You only subtract one time, so choose the subtraction you find easier.

Now multiply the numbers in the circles. Three times 2 is 6. This is the last digit of your answer. The answer is 56.

If a child doesn't know the answer to three times two, I explain it using fingers and hands. I begin by saying that three times two is the same as two times three. We can make two equal the number of hands we have and then three equals the number of fingers we show on each hand. The child then extends three fingers on each hand and counts the number of fingers to get the answer, six.

This is how the completed problem looks:

$7 \times 8 = 56$
③   ②

If you didn't know the numbers to write below 7 and 8 (how many more to make 10), then simply count on your fingers. To find the number to write below 7, you say, 'seven', and then count, 'eight, nine, ten'. You have counted three fingers so write 3 below the 7. There are only five combinations of numbers that add to 10 so you will learn them quickly. Most children have learnt them by the second day.

Let's try 8×9:

$8 \times 9 =$

How many more to make 10 for these two numbers? The answer is 2 and 1. Write 2 and 1 in the circles below the numbers:

$8 \times 9 =$
② ①

What do you do now? You subtract diagonally:

$8 - 1 = 7$ or $9 - 2 = 7$

Seven is the first digit of your answer. Write it down. Now multiply the two circled numbers:

$2 \times 1 = 2$

Two is the last digit of the answer. The answer is 72. Isn't that easy?

If necessary, you can explain that the child can calculate two times one using hands and fingers. Two hands with one finger on each hand gives two fingers or one hand with two fingers will also give an answer of two.

Many younger children have trouble multiplying by one. They often add instead of multiply. If you tell them that one always equals the number of hands and the other number always equals the number of fingers, they should find it easy. One hand and four fingers means 1 times 4 is 4.

Try these

Here are some problems to try by yourself.

(a) $9 \times 9 =$ (e) $8 \times 9 =$
       ◯ ◯   ◯ ◯

(b) $8 \times 8 =$ (f) $9 \times 6 =$
       ② ②   ① ⑩

(c) $7 \times 7 =$ (g) $5 \times 9 =$
       ◯ ◯   ◯ ◯

(d) $7 \times 9 =$ 63   (h) $8 \times 7 =$
       ◯ ◯   ◯ ◯

Do all of the problems, even if you know your tables well. This is the basic strategy we will use for almost all of our multiplication.

How did you go? The answers are:

(a) 81   (b) 64   (c) 49   (d) 63

(e) 72   (f) 54   (g) 45   (h) 56

Isn't this the easiest way to learn your tables?

Now go back and do them all again, but this time do them all in your head. For instance, to multiply the first problem, 9 times 9, you would say, 1 more to make 10. One from 9 is 8, but you wouldn't say 'eight', you would say, 'Eighty ... '.

Then, 1 times 1 is 1. So you say, 'Eighty...one'. Now try the second problem, 8 times 8.

Two more to make 10. Two from 8 or 8 minus 2 is 6. Sixty. Two times 2 is 4. Sixty-four.

Now do the rest by yourself. That was easy, wasn't it?

Multiplying numbers greater than 10

Does this method work for multiplying larger numbers? It certainly does. Let's try it for 96×97:

$$96 \times 97 =$$

What do we take these numbers up to? How many more to make what? One hundred. Ninety-six is 4 below 100 and 97 is 3 below, so we write 4 below 96 and 3 below 97:

$$96 \times 97 =$$
$$\textcircled{4} \quad \textcircled{3}$$

What do we do now? We subtract diagonally. Ninety-six minus 3 or 97 minus 4 equals 93. Write that down as the first part of your answer. What do we do next? Multiply the numbers in the circles. Four times 3 equals 12. Write this down for the last part of the answer. The full answer is 9312:

$$96 \times 97 = 9312$$
$$\textcircled{4} \quad \textcircled{3}$$

Which method is easier, this method or the method you learnt in school? This method, definitely, don't you agree?

Remember my first rule of mathematics:

The easier the method you use to solve a problem, the faster you will solve it with less chance of making a mistake.

 Try these

Here are some more problems to do by yourself.

(a) $96 \times 96 =$ ◯ ◯

(b) $97 \times 95 =$ ◯ ◯

(c) $95 \times 95 =$ ◯ ◯

(d) $98 \times 95 =$ ◯ ◯

(e) $98 \times 94 =$ ◯ ◯

(f) $97 \times 94 =$ ◯ ◯

(g) $98 \times 92 =$ ◯ ◯

(h) $97 \times 93 =$ ◯ ◯

The answers are:

(a) 9216   (b) 9215   (c) 9025   (d) 9310

(e) 9212   (f) 9118   (g) 9016   (h) 9021

Tip

Instead of writing the answers in the book, you may prefer to write the answers on a piece of paper or in an exercise book so you can do the problems again if you wish.

Did you get them all right? If you made a mistake, go back and find where you went wrong and try again. Because the method is so different from what you may have been taught, it is not uncommon to make mistakes at the beginning.

Racing the calculator

I have been interviewed on television programs and in documentaries where the interviewer has asked me to compete with a calculator. This is usually what happens—there is a hand holding a calculator in front of the camera and me in the background. Someone off-screen will call out a problem like 96 times 97. As they call out 96, I immediately take it from 100 and get 4. As they call the second number, 97, I take 4 from it and get an answer of 93. I don't say 93, I say 'Nine thousand, three hundred and...'. I say this with a slow Australian drawl. While I am saying nine thousand, three hundred, I am saying in my mind, 4 times 3 is 12.

So, with hardly a pause I call out, 'Nine thousand, three hundred and...twelve'. Although I don't call myself a lightning calculator—many of my students can beat me—I still have no problem calling out the answer before they get the answer on their calculator.

Now, do the exercises on page 14 again, but this time, do all the calculations in your head. You will find it much easier than you imagine. I tell students they need to do three or four calculations in their head before it really becomes easy—you will find the next time is so much easier than the first. Try it five times before you give up and say it is too difficult.

After many radio interviews, people have telephoned or written to say the method won't work for all numbers. I tell them that it is only because I haven't explained the whole method on air. The method is based on a valid mathematical formula and will work with *any* numbers. In the following chapters I will show you how to apply this method to multiply any numbers.

Are you excited about what you are doing? This is the fast, easy way to master your tables.

Key points

- The method works for multiplying large and small numbers.
- The easier the method you use, the faster you will solve the problem with less chance of making a mistake.

Chapter 3
Using a reference number

We haven't quite reached the end of the explanation for multiplication. The method has worked for all the problems we have done until now, but with a slight adjustment, we can make it work for any numbers.

Using 10 as a reference number

Let's go back to 7 × 8:

⑩ 7 × 8 =
   ◯ ◯

The 10 to the left of the problem is our reference number. It is the number we take our multipliers away from. We ask ourselves, how many do we need to take the number we are multiplying up to the reference number?

We write the reference number to the left of the problem. We then ask ourselves, are the numbers we are multiplying above or below the reference number? In this case the numbers are below, so we put the circles below. How much below the reference number are they? Three and 2. We write 3 and 2 in the circles. Seven is

10 minus 3, so put a minus sign in front of the 3. Eight is 10 minus 2, so put a minus sign in front of the 2.

⑩   7 × 8 =
      −③ −②

We now work diagonally. Seven minus 2 or 8 minus 3 is 5. We write 5 after the equals sign. Now, multiply the 5 by the reference number, 10. How do we multiply by 10? To multiply any number by 10 you simply write a zero after the number. Five times 10 is 50, so write a 0 after the 5. Fifty is our subtotal.

Now multiply the numbers in the circles: 3 times 2 is 6. Add this to the subtotal of 50 for the final answer of 56.

Your full working should look like this:

⑩   7 × 8 = 50
     −③ −② +6
                56   **Answer**

Using 100 as a reference number

In chapter 2 we multiplied 96 × 97. What was the reference number we used? One hundred, because we asked how many more do we need to make 100. The problem worked out in full would look like this:

⑩⓪   96 × 97 = 9300
      −④ −③ +12
             9312   **Answer**

Why do we need to use a reference number?

We need to use this method for multiplying numbers like 6×7 and 6×6.

Let's multiply 6 times 7 and you will see what I mean. We will make the calculation like we did in the previous chapter.

$$6 \times 7 =$$
$$-\text{④} \quad -\text{③}$$

We subtract crossways: $6 - 3 = 3$

$$6 \times 7 = 3$$
$$-\text{④} \quad -\text{③}$$

Now multiply the numbers in the circles.

$$4 \times 3 = 12$$
$$6 \times 7 = 312$$
$$-\text{④} \quad -\text{③}$$

Is the answer correct? Obviously not.

Let's do the calculation in our heads the way we did in the previous chapter.

We put 4 and 3 in the circles.

Six minus 3 is 3. But we don't say three: we say, 'Thirty...'

Then multiplying the numbers in the circles, 4 times 3, we get an answer of 12. So, we say, 'Thirty-twelve'.

However, obviously there is no such number as thirty-twelve. This tells us we have to add the 12 to the first answer of 30 to get an answer of 42.

Written in full with the reference number, our calculation looks like this:

$$
\begin{array}{ccc}
\textcircled{\scriptsize 10} & 6 \times 7 = & 30 \\
-\textcircled{\scriptsize 4} \ -\textcircled{\scriptsize 3} & & \underline{+\ 12} \\
& & 42
\end{array}
$$

The method I explained for doing the calculations in your head actually makes you use this method. Let's multiply 98 by 98 and you will see what I mean.

We take 98 and 98 from 100 and get answers of 2 and 2. We take 2 from 98 and get an answer of 96. But, we don't say 96. We say, 'Nine thousand, six hundred and...'. Nine thousand, six hundred is the answer we get when we multiply 96 by the reference number of 100. We now multiply the numbers in the circles. Two times 2 is 4, so we can say the full answer of nine thousand, six hundred and four.

This is quite impressive because you should be able to give lightning-fast answers to these kinds of problems. You should be able to multiply numbers below 10 very quickly. For example, if you wanted to work out 9 × 9, you would immediately 'see' 1 and 1 below the nines. One from 9 is 8. You would call this 80 (8 times 10). One times 1 is 1. Your answer is 81.

Try these

Try to do these problems in your head.

(a) $96 \times 96 =$

(b) $97 \times 97 =$

(c) $99 \times 99 =$

(d) $95 \times 95 =$

(e) $98 \times 98 =$

Your answers should be:

(a) 9216   (b) 9409   (c) 9801   (d) 9025   (e) 9604

Multiplying numbers in the teens

Let's see how we apply this method to multiplying numbers in the teens. We will use 13×14 as an example and use 10 as our reference number.

⑩ $13 \times 14 =$

Both 13 and 14 are above the reference number, 10, so we put the circles above. How much above are they? Three and 4, so we write 3 and 4 in the circles *above* 13 and 14. Thirteen equals 10 plus 3 so you can write a plus sign in front of the 3; 14 is 10 plus 4 so we can write a plus sign in front of the 4.

$$+\;③\quad +\;④$$
$$⑩\qquad 13 \times 14 =$$

As before, we work diagonally, or crossways. Thirteen plus 4 or 14 plus 3 is 17. Write 17 after the equals sign. We multiply the 17 by the reference number, 10, and get 170. (To multiply any number by 10 we affix a zero.) One hundred and seventy is our subtotal, so write 170 after the equals sign:

$$+\;③\quad +\;④$$
$$⑩\qquad 13 \times 14 = 170$$

For the last step, we multiply the numbers in the circles. Three times 4 equals 12. Add 12 to 170 and we get an answer of 182. This is how we write the problem in full:

$$+\;③\quad +\;④$$
$$⑩\qquad 13 \times 14 = \;170$$
$$+\;12$$
$$\overline{\quad\;\;182\quad}\;\textbf{Answer}$$

If the number we are multiplying is above the reference number we put the circle above. If the number is below the reference number we put the circle below. If the circled number is above we add diagonally, if the number is below we subtract diagonally.

Try these

Try the following problems by yourself.

(a) 12 × 15 =

(b) 13 × 15 =

(c) 12 × 12 =

(d) 13 × 13 =

(e) 12 × 14 =

(f) 12 × 16 =

(g) 14 × 14 =

(h) 15 × 15 =

(i) 12 × 18 =

(j) 16 × 14 =

The answers are:

(a) 180   (b) 195   (c) 144   (d) 169   (e) 168

(f) 192   (g) 196   (h) 225   (i) 216   (j) 224

Tip

If you got any wrong, read through this section again and find your mistake, then try again.

How would you multiply 12 × 21? Let's try it:

$$+②\quad +⑪$$
$$⑩\quad 12 \times 21 =$$

We use a reference number of 10. Both numbers are above 10 so we put the circles above. Twelve is 2 above 10, 21 is 11 above so we write 2 and 11 in the circles. Twenty-one plus 2 is 23, times 10 is 230. Two times 11 is 22, added to 230 makes 252. This is how your completed problem should look:

$$+②\quad +⑪$$
$$⑩\quad 12 \times 21 = \quad 230$$
$$+\ 22$$
$$\overline{252}\quad \textbf{Answer}$$

Solving problems in your head

When we use these strategies, what we say inside our head is very important and can help us solve problems more quickly and easily.

Let's multiply 16 × 16 and look at what we would say inside our head. This is how I would solve the problem—16 plus 6 (from the second 16) equals 22, times 10 equals

220. Six times 6 is 36. Add the 30 first, then the 6. Two hundred and twenty plus 30 is 250, plus 6 is 256.

Inside your head you would say, 'Sixteen plus six, twenty two, two twenty. Thirty six, two fifty six'. With practice, you can leave out half of that. You don't have to give yourself a running commentary on everything you do. You would only say, 'Two twenty, two fifty six'.

Practise doing this.

Tip

Saying the right things in your head as you do the calculations can more than halve the time it takes to do a calculation.

How would you calculate 7×8 in your head? You would 'see' 3 and 2 below the 7 and 8. You would take 2 from the 7 (or 3 from the 8) and say 'Fifty', multiplying by 10 in the same step.

Three times 2 is 'Six'. All you would say is, 'Fifty... six'.

What about 6×7?

You would 'see' 4 and 3 below the 6 and 7. Six minus 3 is 3; you say, 'Thirty'. Four times 3 is 12, plus 30 is 42. You would just say, 'Thirty, forty two'.

It's not as hard as it sounds, is it? And it will become easier the more problems you do.

Does this method replace learning your tables? No, it replaces the method of learning your tables. After you have calculated 7 times 8 equals 56 or 13 times 14 equals 182 a dozen times or more, you stop doing the calculation; you remember the answer. This is much more enjoyable than chanting your tables over and over until they stick.

Key points

- We can use 100 as a reference number to multiply larger numbers.

- With practice, you'll be able to solve large multiplications in your head.

Chapter 4
Multiplying numbers above and below the reference number

Up until now, we have multiplied numbers that were both below the reference number or both above the reference number. How do we multiply numbers when one number is above the reference number and the other is below the reference number?

We will see how this works by multiplying 98 × 135. We will use 100 as our reference number:

(100)   98 × 135 =

Ninety-eight is below the reference number, 100, so we put the circle below. How much below? Two, so we write 2 in the circle. One hundred and thirty-five is above so we put the circle above. How much above? Thirty-five, so we write 35 in the circle above.

$$+ \textcircled{35}$$
$$\textcircled{100} \quad 98 \times 135 =$$
$$- \textcircled{2}$$

One hundred and thirty-five is 100 plus 35 so we put a plus sign in front of the 35. Ninety-eight is 100 minus 2 so we put a minus sign in front of the 2.

We now calculate diagonally. Either 98 plus 35 or 135 minus 2. Then 135 minus 2 equals 133. We write that down after the equals sign. We now multiply 133 by the reference number, 100. One hundred and thirty-three times 100 is 13,300. (To multiply any number by 100, we simply put two zeros after the number.) This is how your work should look at this point:

$$+ \textcircled{35}$$
$$\textcircled{100} \quad 98 \times 135 = 13\,300$$
$$- \textcircled{2}$$

We now multiply the numbers in the circles. Two times 35 equals 70. But that is not really the problem. We have to multiply 35 by –2. The answer is –70. Now your work should look like this:

$$+ \textcircled{35}$$
$$\textcircled{100} \quad 98 \times 135 = 13\,300 - 70 =$$
$$- \textcircled{2}$$

A short cut for subtraction

Let's take a break from this problem for a moment to have a look at a short cut for the subtractions we are doing. What is the easiest way to subtract 70? Let me ask another question. What is the easiest way to take 9 from 56 in your head?

$$56 - 9 = ?$$

I am sure you got the right answer, but how did you get it? Some would take 6 from 56 to get 50, then take another 3 to make up the 9 they have to take away, and get 47.

Some would take away 10 from 56 and get 46. Then they would add 1 back because they took away 1 too many. This would also give them 47.

Some would do the problem the same way they would using pencil and paper. This way they have to carry and borrow in their heads. This is probably the most difficult way to solve the problem.

Tip

Remember, the easiest way to solve a problem is also the fastest, with the lowest chance of making a mistake.

Most people find the easiest way to subtract 9 is to take away 10, then add 1 to the answer. The easiest way to subtract 8 is to take away 10, then add 2 to the answer. The easiest way to subtract 7 is to take away 10, then add 3 to the answer.

- What is the easiest way to take 90 from a number? Take 100 and give back 10.

- What is the easiest way to take 80 from a number? Take 100 and give back 20.

- What is the easiest way to take 70 from a number? Take 100 and give back 30.

If we go back to the problem we were working on, how do we take 70 from 13 300? Take away 100 and give back 30. Is this easy? Let's try it: 13 300 minus 100? The answer is 13 200. Plus 30? The answer is 13 230.

The completed problem looks like this:

$$+ \textcircled{35}$$
$$\textcircled{100} \quad 98 \times 135 = 13\,300 - 70 = 13\,230 \qquad \textbf{Answer}$$
$$-\textcircled{2} \qquad\qquad\qquad \textcircled{30}$$

With a little practice you should be able to solve problems like these entirely in your head.

Try these

Practise with the following problems:

(a) $98 \times 145 =$ (e) $98 \times 146 =$

(b) $97 \times 125 =$ (f) $9 \times 15 =$

(c) $95 \times 120 =$ (g) $8 \times 12 =$

(d) $96 \times 125 =$ (h) $7 \times 12 =$

How did you go? The answers are:

(a) 14 210   (b) 12 125   (c) 11 400   (d) 12 000

(e) 14 308   (f) 135   (g) 96   (h) 84

Multiplying numbers in the circles

This is the rule for multiplying the numbers in the circles:

When both circles are above the numbers or both circles are below the numbers, we add the answer. When one circle is above and one circle is below, we subtract.

When you multiply two positive (plus) numbers together you get a positive (plus) answer. When you multiply two negative (minus) numbers together you get a positive (plus) answer. When you multiply a positive (plus) times a negative (minus) you get a negative answer.

Would our method work for multiplying 8 × 45? Let's try it.

We choose a reference number of 10. Eight is 2 lower than 10 and 45 is 35 higher than 10.

$$+\text{(35)}$$
$$\text{(10)} \quad 8 \times 45 =$$
$$-\text{(2)}$$

You either take 2 from 45 or add 35 to 8. Two from 45 is 43, times the reference number, 10, is 430. Minus 2 times 35 is –70. To take 70 from 430 we take 100, which equals 330, then give back 30 for a final answer of 360.

$$+\text{(35)}$$
$$\text{(10)} \quad 8 \times 45 = 430 - 70 = 360 \quad \textbf{Answer}$$
$$-\text{(2)} \qquad\qquad\quad \text{(30)}$$

Multiplying lower numbers

I don't use this method to multiply numbers like 8 times 12. I think it is easier to simply say, 'Twelve is ten plus two. Ten times eight is eighty, plus two times eight is sixteen'.

$$80 + 16 = 96$$

How about 8 × 13?

Eight times 10 is 80. Eight times 3 is 24. Add the 20 first, and then add the 4.

$$80 + 20 = 100$$
$$100 + 4 = 104$$

Now let's try using the circles and subtracting at the end.

$$\begin{array}{c} +③ \\ ⑩ \quad 8 \times 13 = \\ -② \end{array}$$

Thirteen minus 2 is 11, times 10 is 110.

Minus 2 times 3 is minus 6. This is how the completed problem looks:

$$\begin{array}{c} +③ \\ ⑩ \quad 8 \times 13 = 110 - 6 = 104 \\ -② \end{array}$$

In this case it was easier to simply multiply 8 by 10 and then by 3, and add the answers. But now you have a choice.

Remember, does this replace learning your tables? No, it replaces the *method* of learning your tables. After you have calculated 7 times 8 equals 56 or 13 times 14 equals 182 a dozen times or more, you stop doing the calculation: you remember the answer. This is much more enjoyable than constantly repeating your tables.

We haven't finished with multiplication yet, but we can take a rest here and practise what we have already covered. If some problems don't seem to work out easily, don't worry; we still have more to work through later in the book.

In the next chapter we will take a look at a simple method for checking our answers.

Key points

- We can use a short cut to make our subtractions easier.

- When both circles are *above* the numbers or both circles are *below* the numbers, we *add* the answer. When one circle is *above* and one circle is *below* we *subtract*.

- This method doesn't replace learning your tables: it replaces the *method* of learning them.

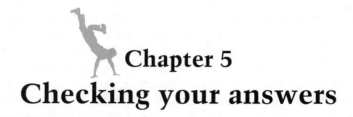

Chapter 5
Checking your answers

How would you like your children to score 100 per cent for every maths test? How would you like them to have a reputation for never making a mistake? In this chapter I will show you how to find and correct mistakes before anyone (including you or their teacher) knows anything about them.

I often tell my students it is not enough to calculate an answer to a problem in mathematics: you haven't finished until you have checked you have the right answer.

The following method of checking answers has been known to mathematicians for about a thousand years, but it doesn't seem to have been taken seriously by educators in most countries.

How do I check my work?
When I was young, I used to make a lot of careless mistakes in my calculations. I knew how to do the problems, I would do everything the right way, but I still got the wrong answer. I would forget to carry a number, or say the right answer but write down something different, and who knows what other mistakes I made?

My teachers and my parents would tell me to check my work. The only way I knew to check my work was to do the problem again. But if I got a different answer, when did I make the mistake? Maybe I got it right the first time and made a mistake the second time. So I would have to solve the problem a third time. If two answers agreed, that was probably the right answer. But maybe I made the same mistake twice. So my teachers would tell me to try to solve the problem in two different ways, which was good advice, but they didn't give me time in my math tests to do each question three times. Had someone taught me what I am about to teach you, I could have had a reputation for being a mathematical genius.

I am disappointed that this method was known, but nobody taught it. It is called the digit sum method, or casting out nines. This is how it works.

Substitute numbers

To check the answer to a calculation, we use substitute numbers instead of the original numbers we were working with. A substitute on a football or basketball team is somebody who replaces somebody else on the team: they take another person's place. That's what we do with the numbers. We use substitute numbers in place of the original numbers to check our work.

Let's try an example. Say we have just calculated 13×14 and got an answer of 182. We want to check our answer.

$13 \times 14 = 182$

The first number in the problem is 13. We add its digits together to get the substitute:

$1 + 3 = 4$

Four is our substitute for 13. The next number we work with is 14. We add its digits:

$$1 + 4 = 5$$

Five is our substitute for 14. We now do the same calculation (multiplication) using the substitute numbers instead of the original numbers:

$$4 \times 5 = 20$$

Twenty is a two-digit number so we add its digits together to get our check answer:

$$2 + 0 = 2$$

Two is our check answer.

 Tip

If we have the right answer in our original calculation, the digits in the original answer should add up to the same as the digits in our check answer.

We add the digits of the original answer, 182:

$$1 + 8 + 2 = 11$$

Eleven is a two-digit number so we add its digits together to get a one-digit answer:

$$1 + 1 = 2$$

Two is our substitute answer. This is the same as our check answer, so our original answer is correct.

Let's try it again, this time using 13×15:

$$13 \times 15 = 195$$

$$1 + 3 = 4 \text{ (substitute for 13)}$$

$$1 + 5 = 6 \text{ (substitute for 15)}$$

So our substitute numbers are 4 and 6. The next step is to multiply these:

$$4 \times 6 = 24$$

Twenty-four is a two-digit number so we add its digits:

$$2 + 4 = 6$$

Six is our check answer.

Now, to find out if we have the correct answer, we check this by adding the digits in our original answer, 195:

$$1 + 9 + 5 = 15$$

To bring 15 to a one-digit number:

$$1 + 5 = 6$$

Six is what we got for our check answer so we can be confident we didn't make a mistake.

A short cut

There is a short cut to this procedure. If we find a 9 anywhere in the calculation, we cross it out. Using the previous example, you can see how this removes a step from our calculations without affecting the result. With the last answer, 195, instead of adding $1 + 9 + 5$, which equals 15, and then adding $1 + 5$, which equals 6, we could cross out the 9 and just add 1 and 5, which also equals 6. This makes no difference to the answer, but it saves some time and effort, and I am in favour of anything that saves time and effort. That is why this method of checking is called 'casting out nines'.

What about the answer to the first problem we solved, 182? Can we use this short cut?

We added $1 + 8 + 2$ to get 11, then added $1 + 1$ to get our final check answer of 2. In 182, there are two digits that add up to 9, the 1 and the 8. Cross them out and you just have the 2 left. No more work to do at all, so the short cut works.

Let's try it again for more of an idea of how it works:

$$167 \times 346 = 57\,782$$

To find our substitute for 167:

$$1 + 6 + 7 = 14$$

$$1 + 4 = 5$$

There were no short cuts with the first number. Five is our substitute for 167.

To find our substitute for 346:

$$3 + 4 + 6 =$$

We immediately see that $3 + 6 = 9$, so we cross out the 3 and the 6. That just leaves us with 4, our substitute for 346.

Can we find any nines, or digits adding up to 9 in the answer? Yes, $7 + 2 = 9$, so we cross out the 7 and the 2. We add the other digits:

$5 + 7 + 8 = 20$

$2 + 0 = 2$

Two is our substitute answer.

I write the substitute numbers in pencil above or below the actual numbers in the problem. It might look like this:

$167 \times 346 = 57782$

   5   4   2

Is 57 782 the right answer?

We multiply the substitute numbers, 5 times 4 equals 20. The digits in 20 add up to 2 $(2 + 0 = 2)$. This is the same as our substitute answer so we were right again.

Let's try one more example. Use this method to check if this answer is correct:

$456 \times 831 = 368\,936$

We write down our substitute numbers:

$456 \times 831 = 368936$

   6   3   8

That was easy because we cast out (or crossed out) 4 and 5 from the first number, leaving 6; we cast out 8 and 1 from the second number, leaving 3, and almost every digit was cast out of the answer, 3 plus 6 twice, and a 9, leaving a substitute answer of 8.

We now see if the substitutes work out correctly. Six times 3 is 18, which adds up to 9, which is also cast out, leaving zero. But our substitute answer is 8, so we made a mistake somewhere.

When we calculate it again, we get 378 936.

Did we get it right this time? The 936 cancels out, so we add 3 + 7 + 8, which equals 18, and 1 + 8 adds up to 9, which cancels, leaving zero. This is the same as our check answer, so this time we have it right.

Simple and fast

Does this short cut prove that we have the right answer? No, but we can be almost certain. For instance, say we got 3 789 360 for our last answer—by mistake we put a zero on the end. The final zero wouldn't affect our check by casting out nines and we wouldn't know we had made a mistake. When it showed we had made a mistake, though, the check definitely proved that we had the wrong answer. It is a simple, fast check that will find most mistakes, and it should help your child get 100 per cent for most of his or her math tests.

Why does the method work?

You will be much more successful using a new method when you know not only that it does work, but understand why it works as well. So read on as I explain why casting out the nines works.

Think of a number and multiply it by 9. What is 4 × 9? The answer is 36. Add the digits in the answer together, 3 + 6, and you get 9.

Let's try another number. Three nines are 27. Add the digits of the answer together, 2 + 7, and you get 9 again.

Eleven nines are 99. Nine plus 9 equals 18. Wrong answer? No, not yet. Eighteen is a two-digit number so we add its digits together—1 + 8. Again, the answer is 9.

If you multiply any number by nine, the sum of the digits in the answer will always add up to nine if you keep adding the digits until you get a one-digit number. This is an easy way to tell if a number is evenly divisible by nine. If the digits of any number add up to nine, or a multiple of nine, then the number itself is evenly divisible by nine.

That is why, when you multiply any number by nine, or a multiple of nine, the digits of the answer must add up to nine. For instance, what if you were checking the following calculation:

$$135 \times 83\,615 = 11\,288\,025$$

Add the digits in the first number:

$$1 + 3 + 5 = 9$$

To check our answer, we don't need to add the digits of the second number, 83 615, because we know 135 has a digit sum of 9, so if our answer is correct it, too, should have a digit sum of 9.

Let's add the digits in the answer:

$$1 + 1 + 2 + 8 + 8 + 0 + 2 + 5 =$$

Eight plus 1 cancels twice, leaving $2 + 2 + 5 = 9$, so we were right.

You can have fun playing with this.

If the digits of a number add up to any number other than nine, this other number is the remainder you would get after dividing the number by nine.

Let's take 14:

$$1 + 4 = 5$$

Five is the digit sum of 14. It should be the remainder you would get if you divided by 9. Nine goes into 14 once, with 5 remainder. If you add 3 to the number, you add 3 to the remainder. If you double the number, you double the remainder. Whatever you do to the number, you do to the remainder, so we can use the remainders as substitutes.

Why do we use 9 remainders; couldn't we use the remainders after dividing by, say, 17? Certainly, but there is so much work involved in dividing by 17, the check would be harder than the original problem. We choose nine because of the easy short cut for finding the remainder.

 Tip

Remember, the easiest way to solve a problem is also the fastest, with the least chance of making a mistake.

Key points

- Always check the answer when you have completed a problem.

- Using substitute numbers is a quick and accurate way to check your work.

Chapter 6
Multiplication — more about the methods

In chapters 3 to 5 we multiplied numbers using an easy method that makes multiplication fun. It is easy to use when the numbers are near 10 or 100. But what about multiplying numbers that are around 30 or 60? Can we still use this method? We certainly can.

We choose reference numbers of 10 and 100 because it is easy to multiply by those numbers. The method will work just as well with other reference numbers, but we must choose numbers that are simple to multiply by.

Multiplication by factors

It is easy to multiply by 20, as 20 is two times ten. It is easy to multiply by 10 and it is easy to multiply by two. This is called multiplication by factors, as 10 and two are factors of 20 $(10 \times 2 = 20)$.

Let's try an example:

$23 \times 24 =$

Twenty-three and 24 are above the reference number, 20, so we put the circles above. How much above are they?

The answer is 3 and 4. We write those numbers above in the circles. We write them above because they are plus numbers (23 = 20 + 3, 24 = 20 + 4).

$$+\textcircled{3} \quad +\textcircled{4}$$
$$\textcircled{20} \qquad 23 \times 24 =$$

We add diagonally as before:

$$23 + 4 = 27$$

We now multiply this answer, 27, by the reference number 20. To do this we multiply by 2, then by 10:

$$27 \times 2 = 54$$
$$54 \times 10 = 540$$

(The easy way to multiply 27 by 2 is to see 27 as 30 − 3. Twice 30 − 3 is 60 − 6, which is 54.)

The rest is the same as before. We multiply the numbers in the circles:

$$3 \times 4 = 12$$
$$540 + 12 = 552$$

Your work would look like this:

$$+\textcircled{3} \quad +\textcircled{4}$$
$$\textcircled{20} \qquad 23 \times 24 = \quad 27$$
$$540$$
$$+ \, 12$$
$$\overline{552} \quad \textbf{Answer}$$

Doubling numbers

It is easy to double numbers when you are using 20 as a reference number. If you are multiplying 14 by 2 you just double the 1 and the 4 to get 28. If you want to double 28 there is an easy method. Look at 28 and see it as 30 minus 2. Double 30 minus 2 is 60 minus 4, or 56. When you have a high units digit, instead of doubling it and carrying it is easier to go up to the next tens. So, you would see 37 as 40 − 3 and 19 as 20 − 1. That makes the calculation easier to do in your head.

 Try these

How would you double the following numbers?

(a) 29   (b) 18   (c) 47   (d) 36

The answers are:

(a) $(30 − 1) \times 2 = 60 − 2 = 58$

(b) $(20 − 2) \times 2 = 40 − 4 = 36$

(c) $(50 − 3) \times 2 = 100 − 6 = 94$

(d) $(40 − 4) \times 2 = 80 − 8 = 72$

 Tip

Try to make a habit of using the easiest method for any calculation.

Checking your answers

Let's apply what we covered in the previous chapter and check our answer:

$$23 \times 24 = 552$$
$$5 \quad 6 \quad 12$$
$$3$$

The substitute numbers for 23 and 24 are 5 and 6.

$$5 \times 6 = 30$$
$$3 + 0 = 3$$

Three is our check answer.

The digits in our original answer, 552, add up to 3:

$$5 + 5 + 2 = 12$$
$$1 + 2 = 3$$

This is the same as our check answer, so we were right. Let's try another:

$$23 \times 31 =$$

We put 3 and 11 above 23 and 31:

$$+③ \quad +⑪$$
$$⑳ \quad 23 \times 31 =$$

They are 3 and 11 above the reference number, 20. Adding diagonally, we get 34:

$31 + 3 = 34$ or $23 + 11 = 34$

We multiply this answer by the reference number, 20. To do this, we multiply 34 by 2, then multiply by 10.

$34 \times 2 = 68$

$68 \times 10 = 680$

This is our subtotal. We now multiply the numbers in the circles (3 and 11):

$3 \times 11 = 33$

Add this to 680:

$680 + 33 = 713$

The calculation will look like this:

$$
\begin{array}{r}
^{+\text{\textcircled{3}}\ \ +\text{\textcircled{11}}} \\
\text{\textcircled{20}} \quad 23 \times 31 = \quad 34 \\
680 \\
+\ 33 \\
\hline
713 \quad \textbf{Answer}
\end{array}
$$

We check by casting out the nines:

$23 \times 31 = 713$

5   4   11

2

Multiply our substitute numbers and then add the digits in the answer:

$$5 \times 4 = 20$$

$$2 + 0 = 2$$

This checks with our substitute answer so we can accept that as correct.

 Try these

Here are some problems to try for yourself. When you have finished them, check your answers by casting out the nines.

(a) 21 × 26 =

(d) 23 × 27 =

(b) 24 × 24 =

(e) 21 × 36 =

(c) 23 × 23 =

(f) 26 × 24 =

You should be able to do all of those problems in your head. It's not difficult with a little practice.

Multiplying numbers below 20

How about multiplying numbers below twenty? If the numbers (or one of the numbers to be multiplied) are in the high teens, we can use 20 as a reference number.

Let's try an example:

$$19 \times 16 =$$

Using 20 as a reference number we get:

⑳ $19 \times 16 =$
    $-①$ $-④$

Subtract diagonally:

$16 - 1 = 15$

Multiply by 20:

$2 \times 15 = 30$

$30 \times 10 = 300$

Our subtotal is 300.

Now we multiply the numbers in the circles and then add the result to our subtotal:

$1 \times 4 = 4$

$300 + 4 = 304$

Your completed work should look like this:

⑳ $19 \times 16 =$ 15
    $-①$ $-④$ 300
                  $\underline{+\ 4}$
                  304   **Answer**

Now let's try the same example using 10 as a reference number:

$$+\textcircled{9} \quad +\textcircled{6}$$
$$\textcircled{10} \qquad 19 \times 16 =$$

Add diagonally, then multiply by 10 to get a subtotal:

$$19 + 6 = 25$$

$$10 \times 25 = 250$$

Multiply the numbers in the circles and add this to our subtotal:

$$9 \times 6 = 54$$

$$250 + 54 = 304$$

Your completed work should look like this:

$$+\textcircled{9} \quad +\textcircled{6}$$
$$\textcircled{10} \qquad 19 \times 16 = \quad 250$$
$$+ 54$$
$$\overline{\quad 304 \quad} \quad \textbf{Answer}$$

This confirms our first answer. There isn't much to choose between using the different reference numbers. It is a matter of personal preference. Simply choose the reference number you find easier to work with.

The third possibility is if one number is above and the other is below 20.

$$+ ⑫$$
$$⑳ \quad 18 \times 32 =$$
$$- ②$$

We can either add 18 to 12 or subtract 2 from 32, and then multiply the result by our reference number:

$$32 - 2 = 30$$

$$30 \times 20 = 600$$

We now multiply the numbers in the circles:

$$2 \times 12 = 24$$

It is actually −2 times 12 so our answer is −24.

$$600 - 24 = 576$$

Your work should now look like this:

$$+ ⑫$$
$$⑳ \quad 18 \times 32 = 30$$
$$- ② \qquad\qquad 600 - 24 = 576$$
$$⑥$$

Let's check the answer by casting out the nines.

$$18 \times 32 = 576$$

9	5	18
0		0

Zero times 5 is 0, so the answer is correct.

Multiplying higher numbers

That takes care of the numbers up to around 30 times 30. However, what do we do if the numbers are higher? We can use 50 as a reference number. It is easy to multiply by 50 because 50 is half of 100, or 100 divided by two. So, to multiply by 50, we multiply the number by 100, and then divide that answer by two.

Let's try it:

$$\text{\textcircled{50}} \quad 46 \times 48 =$$
$$-\text{\textcircled{4}} \quad -\text{\textcircled{2}}$$

Subtract diagonally:

$$46 - 2 = 44$$

Multiply 44 by 100:

$$44 \times 100 = 4400$$

To say it in your head, just say, 'Forty-four by one hundred is forty-four hundred'. Then halve, to multiply by 50, which gives you 2200.

$4400 \div 2 = 2200$

Then multiply the numbers in the circles, and add this result to 2200:

$4 \times 2 = 8$

$2200 + 8 = 2208$

Your completed work should look like this:

$\underset{-④\ -②}{⑤⓪}\quad 46 \times 48 = 4400$

$$\begin{array}{r} 2200 \\ +\ 8 \\ \hline 2208 \quad \textbf{Answer} \end{array}$$

That was so easy. Let's try another:

$\underset{⑤⓪}{\overset{+③\ +⑦}{53 \times 57 =}}$

Add diagonally, then multiply the result by the reference number (multiply by 100 and then divide by 2):

$57 + 3 = 60$

$60 \times 100 = 6000$

Divided by 2 equals 3000.

Then multiply the numbers in the circles and add the result to 3000:

$3 \times 7 = 21$

$3000 + 21 = 3021$

Your work should now look like this:

$$+\text{③} \quad +\text{⑦}$$
$$\text{㊿} \qquad 53 \times 57 = 6000$$

$$\begin{array}{r} 3000 \\ + 21 \\ \hline 3021 \end{array} \quad \textbf{Answer}$$

Let's try one more:

$$+\text{②} \quad +\text{⑬}$$
$$\text{㊿} \qquad 52 \times 63 =$$

Add diagonally and multiply the result by the reference number (multiply by 100 and then divide by 2):

$63 + 2 = 65$

$65 \times 100 = 6500$

Then we halve the answer. If we say six thousand five hundred, the answer is easy. Half of 6000 is 3000. Half of 500 is 250. Our subtotal is 3250.

Now multiply the numbers in the circles:

$2 \times 13 = 26$

Add 26 to our subtotal and we get 3276. Your work should now look like this:

+② +③

㊿   52 × 63 = 6500
                    3250
                  + 26
                  ─────
                    3276   **Answer**

We could check that by casting out the nines:

52 × 63 = 3276

7   0   0

Six plus 3 in 63 add up to 9, which cancels to leave 0. In the answer, $3 + 6 = 9$, $2 + 7 = 9$, it all cancels. Seven times zero does give us zero, so the answer is correct.

Here are some problems for you to do. See how many you can do in your head.

 Try these

(a) 46 × 42 =   (e) 51 × 55 =

(b) 47 × 49 =   (f) 54 × 56 =

(c) 46 × 47 =   (g) 51 × 68 =

(d) 44 × 44 =   (h) 51 × 72 =

How did you go with those? You should have had no trouble doing all of them in your head. Now check your answers by casting out the nines.

Multiplying lower numbers

Let's look at multiplying numbers that don't work well with a reference number of 10. We'll try 6 times 4 as an example.

$$\overset{\text{\tiny(10)}}{} \quad 6 \times 4 =$$
$$\quad -\text{\small(4)} \ -\text{\small(6)}$$

We use a reference number of 10. The circles go below because the numbers 6 and 4 are lower than 10. We subtract diagonally:

$$6 - 6 = 0 \quad \text{or} \quad 4 - 4 = 0$$

We then multiply the numbers in the circles:

$$4 \times 6 =$$

That was our original problem. The method doesn't seem to help. Can we make our method or formula work in this case? We can, but we must use a different reference number. Let's try a reference number of five. Five is 10 divided by two, or half of ten. The easy way to multiply by five is to multiply by ten and halve the answer.

$$+\text{\small(1)}$$
$$\text{\small(5)} \quad 6 \times 4 =$$
$$-\text{\small(1)}$$

Six is above 5 so we put the circle above. Four is below 5 so we put the circle below. Six is 1 higher than 5 and 4 is 1 lower, so we put 1 in each circle.

We add or subtract diagonally:

$$6 - 1 = 5 \quad \text{or} \quad 4 + 1 = 5$$

We multiply 5 by the reference number, which is also 5.

To do this, we multiply by 10, which gives us 50, and then divide by 2, which gives us 25. Now we multiply the numbers in the circles:

$$1 \times -1 = -1$$

Because the result is a negative number, we subtract it from our subtotal rather than adding it:

$$25 - 1 = 24$$

This is how the completed problem looks:

$$
\begin{array}{c}
+\,\textcircled{1} \\
\textcircled{5} \quad 6 \times 4 = 5 \\
-\,\textcircled{1} \quad 25 - 1 = 24 \quad \textbf{Answer}
\end{array}
$$

This is a long-winded and complicated method for multiplying low numbers, but it shows we can make this method work with a little ingenuity. Actually, these strategies will develop our ability to think laterally, which is very important for mathematicians and also for succeeding in life.

Let's try some more, even though you probably know your lower tables quite well:

⑤   4 × 4 =
      –① –①

Subtract diagonally:

$4 - 1 = 3$

Multiply your answer by the reference number:

$3 \times 10 = 30$

Thirty divided by 2 equals 15. Now multiply the numbers in the circles:

$1 \times 1 = 1$

Add that to our subtotal:

$15 + 1 = 16$

This is how the completed problem looks:

⑤   4 × 4 = 30
      –① –①   15
                    + 1
                    16   **Answer**

Try these

Now try the following:

(a) 3 × 4 =   (d) 3 × 6 =

(b) 3 × 3 =   (e) 3 × 7 =

(c) 6 × 6 =   (f) 4 × 7 =

The answers are:

(a) 12   (b) 9   (c) 36   (d) 18   (e) 21   (f) 28

I'm sure you had no trouble doing those. I don't really think that is the easiest way to learn those tables. I think it is easier to simply remember them. Some people want to learn how to multiply low numbers just to check that the method will work. Others like to know that if they can't remember some of their tables, there is an easy method to calculate the answer. Even if you know your tables for these numbers, it is still fun to play with numbers and experiment.

Multiplying by 5

As we have seen, to multiply by 5 we can multiply by 10 and halve the answer. Five is half of ten. To multiply 6 by 5 we can multiply 6 by 10, which is 60, and then halve the answer to get 30.

Try these

Here are some you can try:

(a) $8 \times 5 =$ (c) $2 \times 5 =$

(b) $4 \times 5 =$ (d) $6 \times 5 =$

The answers are:

(a) 40   (b) 20   (c) 10   (d) 30

This is what we do when the tens digit is odd. Let's try 7×5:

$$7 \times 10 = 70$$

If you find it difficult to halve 70, split it into $60 + 10$. Half of $60 + 10$ is $30 + 5$, which equals 35. Let's try another.

$$9 \times 5 =$$

Ten nines are 90. Ninety splits into $80 + 10$. Half of $80 + 10$ is $40 + 5$, so our answer is 45.

Try these

Try these for yourself:

(a) $3 \times 5 =$

(c) $9 \times 5 =$

(b) $5 \times 5 =$

(d) $7 \times 5 =$

The answers are:

(a) 15   (b) 25   (c) 45   (d) 35

This is an easy way to teach the five times table.

Key points

- Multiplying using a reference number will work for any number.
- Our check (casting out the nines) also works for all numbers.

Chapter 7
A fun mathematical short cut — multiplying by 11

Many books have been written about mathematical short cuts. Many short cuts will not only save time and effort, they can also be useful in developing number skills at the same time. In this chapter we will look at a fun short cut you can play with and use to show off your skills to your friends. It will also help you to understand numbers better.

Multiplication by 11

To multiply a two-digit number by 11, simply add the two digits together and insert the result in between.

For example, to multiply 23 by 11, add 2 plus 3, which equals 5, and insert the 5 between the 2 and the 3, for the answer, 253.

To multiply 14 by 11, add 1 plus 4, which equals 5, and insert the 5 between the 1 and the 4, to give us the answer 154.

Try these

Try these for yourself:

(a) 63 × 11 =   (d) 26 × 11 =

(b) 52 × 11 =   (e) 71 × 11 =

(c) 34 × 11 =   (f) 30 × 11 =

The answers are:

(a) 693   (b) 572   (c) 374

(d) 286   (e) 781   (f) 330

In the examples above, the two digits add up to less than 9. What do we do if the two digits add to a number higher than 9? If the result is a two-digit number, insert the units digit of the result between the digits and carry the one to the first digit of the answer.

For instance, to multiply 28 by 11, add 2 to 8, which equals 10. Insert the 0 between the 2 and 8 to get 208, and carry the one to the first digit, 2, to give an answer of 308.

Let's try 88 × 11:

$$8 + 8 = 16$$

Insert 6 between the 88 to give 868, then add 1 to the first 8 to give an answer of 968.

Calling out the answers

Giving children these problems (and doing them ourselves) will develop basic number skills. Children of all ages will become very quick at calling out the answers.

If someone were to ask you to multiply 77 by 11, you would immediately see that 7 plus 7 equals 14, which is more than 10. You would immediately add 1 to the first 7 and call out, 'Eight hundred and...'. The next digit will be the 4 from the 14, followed by the remaining 7, so then you could say, '...forty...seven'. Try it. It is much easier than it sounds.

Here is another example—if you had to multiply 84 by 11, you would see that 8 plus 4 is more than 9, so you would add 1 to the 8 to give, 'Nine hundred and...' Now we add 8 and 4, which is 12, so the middle digit is two. You would say, '...twenty...'. The final digit remains 4, '...four'. Your answer is, 'Nine hundred and twenty-four'.

How about 96 × 11?

Nine plus 6 is 15. Add one to the 9 to get 10. Work with 10 as you would a single-digit number—10 is the first part of the answer. Five is the middle digit. Six remains the final digit. The answer is 1056.

If you were doing this problem in your head, you would say, 'Nine plus one carried is ten'. Out loud you would say, 'One thousand and...'. Then you would see that the 5 from the 15 is the tens digit, so you continue, '...fifty...'. Then, the units digit remains the same, 6. You would give the answer, 'One thousand and...fifty...six'.

Try these

Try these for yourself. Call out the answers as fast as you can.

(a) $37 \times 11 =$ (d) $92 \times 11 =$

(b) $48 \times 11 =$ (e) $82 \times 11 =$

(c) $76 \times 11 =$ (f) $66 \times 11 =$

The answers are:

(a) 407   (b) 528   (c) 836

(d) 1012   (e) 902   (f) 726

Multiplying larger numbers

To multiply larger numbers by 11, we use a similar method. Let's take the example of $12\,345 \times 11$. We would write the problem like this:

012345×11

We write a zero in front of the number we are multiplying. You will see why in a moment. Beginning with the units digit, add each digit to the digit on its right. In this case, add 5 to the digit on its right. There is no digit on its right, so add nothing:

$5 + 0 = 5$

Write 5 as the last digit of your answer. Your calculation should look like this:

$$\frac{012345 \times 11}{5}$$

Now go to the 4. Five is the digit on the right of the 4:

$$4 + 5 = 9$$

Write 9 as the next digit of your answer. Your calculation should now look like this:

$$\frac{012345 \times 11}{95}$$

Continue the same way:

$$3 + 4 = 7$$
$$2 + 3 = 5$$
$$1 + 2 = 3$$
$$0 + 1 = 1$$

Here is the finished calculation:

$$\frac{012345 \times 11}{135795}$$

If we hadn't written the zero in front of the number to be multiplied, we might have forgotten the final step.

This is an easy method to multiply by 11. The strategy develops addition skills while students are using the method as a short cut.

Let's try another problem. This time we will have to carry digits. The only digit you can carry, using this method, is 1.

Let's try this example:

217 475 × 11

We write the problem like this:

0 217 475 × 11

We add the units digit to the digit on its right. There is no digit to the right of 5, so 5 plus nothing is 5. Write down 5 below the 5. Now add the 7 and the 5:

$7 + 5 = 12$

Write the 2 as the next digit of the answer and carry the 1. Your working should look like this:

$$\frac{0217\,4^{1}75 \times 11}{25}$$

The next steps are:

$4 + 7 + 1 \text{ (carried)} = 12$

Write 2 and carry 1. Add the next numbers:

$7 + 4 + 1 \text{ (carried)} = 12$

Two is again the next digit of the answer and carry 1.

$1 + 7 + 1$ (carried) $= 9$

Write 9. Then:

$2 + 1 = 3$

$0 + 2 = 2$

Here is the finished calculation:

$021^17\,^14^175 \times 11$
——————————
$2\,392\,225$ **Answer**

A maths game

We can also use this method as a game. This involves a simple check for multiplication by 11 problems. The problem isn't completed until we have checked it. Let's check the problem discussed on pages 70–71:

$012\,345 \times 11$
——————
$135\,795$

Write an x under every second digit of the answer, beginning from the right-hand end of the number. The calculation will now look like this:

$012\,345 \times 11$
——————
$135\,795$
  x x x

Add the digits with the x under them:

$1 + 5 + 9 = 15$

Add the digits without the x:

$3 + 7 + 5 = 15$

If the answer is correct, the answers will either be the same, have a difference of 11, or a difference of a multiple of 11, such as 22, 33, 44 or 55. Both added to 15, so our answer is correct. This is a test to see if a number can be evenly divided by 11.

Let's check the problem on pages 72–73:

$$\frac{0\,217\,475 \times 11}{2\,392\,225}$$
$$\text{x}\quad\text{x}\quad\text{x}$$

Add the digits with the x under them:

$3 + 2 + 2 = 7$

Add the digits without the x:

$2 + 9 + 2 + 5 = 18$

To find the difference between 7 and 18, we take the smaller number from the larger number:

$18 - 7 = 11$

If the difference is 0, 11, 22, 33, 44, 55, 66 and so on, then the answer is correct. We have a difference of 11, so our answer is correct.

Give the problem to children. Ask them to make up their own numbers to multiply by 11 and see how big a difference they can get. The larger the number they are multiplying, the greater the difference is likely to be. Let them try for a new record.

Children will multiply a 700-digit number by eleven in their attempt to set a new record. By trying for a new world record, they are improving their basic addition skills and checking their work as they go.

Key points

- Short cuts are fun and easy to learn.
- Using short cuts improves your mental calculation skills.

Chapter 8
Factors

I owe a lot to my third grade and fourth grade teachers, Miss Clark and Mrs O'Connor. They obviously did not teach us according to the education department's guidelines. They introduced us to the concept of using factors to simplify mathematical calculations.

In third grade, Miss Clark taught the class long multiplication, but introduced it by teaching us to use factors. When she taught us standard long multiplication for multiplying by numbers that couldn't be broken down into factors, we realised there was an added step involved—two multiplications plus an addition: after we had multiplied twice we had to add the two answers together. Why would you use standard long multiplication to multiply by a number that could be factorised?

Then in grade four, Mrs O'Connor taught us long division by factors before teaching us standard long division. An understanding of factors will help with standard long division and make the calculation much easier. Many people panic at the idea of long division but using factors makes it easy to multiply and divide by large numbers.

The other grade three and four students at my school missed out on this teaching.

What are factors?

Factors are the numbers you multiply together to get another number. A number may be made by multiplying two or more other numbers together. The numbers that are multiplied together are called factors of the final number.

For instance, you can multiply 2 times 3 to get an answer of 6. Two and three are factors of six. Or, six has factors of two and three. If you want to multiply a number by six you could multiply the number by three and then multiply the answer by two. That would be multiplication by factors.

We used factors in chapter 6 when we used 20 as a reference number. To multiply by 20 we multiplied the number by two and then the answer by 10. To multiply by 15 using factors you would multiply by five and then multiply the answer by three, because 3 times 5 is 15. To multiply by 12 using factors you could multiply by two and by six, or you could multiply by four and by three.

What are factors of 15?

What are factors of 25?

What are factors of 21?

Fifteen has factors of 3 and 5 $(3 \times 5 = 15)$.

Five and 5 are factors of 25, and 3 and 7 are factors of 21.

What is the easy way to multiply 3 by 16? You could say 3 times 10 is 30, plus 3 times 6 is 18. Thirty plus 18 gives us 48. There is an easier way.

What numbers multiplied together give an answer of 16? You could multiply 4 times 4 or 2 times 8. These are factors of 16.

Now to multiply 3 × 16 you could say, 3 times 2 by 8, because 2 times 8 is 16. Three times 2 is easy: 6; and now the calculation is 6 times 8. You probably know the answer but, if not, you use the circles to calculate 6 × 8 = 48.

So, 3 × 16 = 48.

If you know that 3 times 8 is 24 then you could multiply 3 by 8 and then multiply the answer by 2. Three times 8 is 24. It is easy to double 24: 24 doubled (multiplied by 2) is 48.

Or, you could use 4 × 4 as the factors. Three times 4 equals 12; 12 times 4 is 48. You have a choice of methods.

Tip

Miss Clark gave grade three some advice: generally it is easier to multiply by the larger factor first, and then multiply by the smaller factor. That means you have a smaller number to multiply by the larger factor.

Multiplying by multiples of 10

Many students don't realise how easy it is to multiply by a multiple of 10. That is, to multiply a number by 30, 40 or 50. They would find it easy to multiply 3 × 4 = 12 but be lost if they had to multiply 30 × 40 in their heads. To multiply by 30 you just have to remember that thirty is 3 × 10, so you simply multiply by 3 and then add a nought to the answer. Miss Clark told us that the easy way to multiply 30 × 40 was to multiply 3 × 4 and add two zeroes. Keep in mind that 30 × 40 = 3 × 10 × 4 × 10. You can multiply the factors in any order you like so the easy way would be 3 × 4 × 10 × 10. You would multiply 3 × 4 = 12 and then add a nought for each of the tens you had to multiply. Your answer would be 1200.

 Try these

Practise using factors with the following:

(a) 5 × 30 =

(b) 2 × 60 =

(c) 4 × 20 =

The answers are:

(a) 150   (b) 120   (c) 80

Here are some more problems that are very easy if you can see 14 as 2 times 7 or 16 as 2 times 8, or 24 as 2 times 12.

Looking at the next 'Try these' box, let's do (c) together. If you can see 24 as 2 times 12 the problem becomes 12 × 2 × 6. If you multiply 2 × 6 first you get 12 times 12. The answer is 144.

Try these

Here are some more to try:

(a) 4 × 14 =

(b) 5 × 16 =

(c) 24 × 6 =

The answers are:

(a) 56   (b) 80   (c) 144

Maths problem in a novel

I was reading a book recently where, as part of the plot, a teenager had to multiply 8 times 99. He gave the answer in a flash and everyone thought he was extra smart. The problem was that the answer in the book was wrong. It is a fact, though, that if you can calculate quickly in your head then everyone thinks you are highly intelligent. They will think you are a brain.

How do you calculate 8 times 99? You could multiply 8 by 100 and subtract 8. That is not hard to do. Eight times 100 is 800, minus 8 is 792.

I used factors to solve it. Ninety-nine is 9 × 11. Multiply 8 × 9 to get 72. You know the short cut for multiplication by 11. Add 7 plus 2 equals 9 and put the 9 in the middle to get 792. Either way is easy.

Know your tables

To get the most benefit from using factors you need to know your tables well to be able to recognise numbers that can be factorised and to know the factors of the numbers.

Key points

- Practise breaking numbers up into factors.

- Mathematician Wim Klein used to walk through a car park and break number plates into factors.

- When I watch limited-overs cricket, I calculate the run rate by using factors. Apply the method to your own interests.

- Have fun with factors.

Chapter 9
Multiplying using two reference numbers

Multiplying using a reference number has worked well for numbers that are close to each other. When the numbers are not close, the method still works but the calculation is more difficult. For instance, what if we wanted to multiply numbers like 13 × 64? Which reference number would we choose? In this chapter, we will look at an easy way to use the formula with two reference numbers.

Using two reference numbers

It is possible to multiply two numbers that are not close to each other by using two reference numbers. I will work through an example and then I will explain how the method works. I will take 8 × 27 as an example. Eight is close to 10, so we will use 10 as the base reference number. Twenty-seven is close to 30, so we will use 30 as the second reference number. From the two reference numbers, we choose the easiest number to multiply. It is easy to multiply by 10, so we will choose 10. This becomes the base reference number. The second reference number should be a multiple of the base reference number. The number we have chosen, 30, is 3 times the base reference number, 10. Instead of a circle, I write the two reference numbers to the left of the problem in brackets.

The base reference number is 10. The second reference number is 30, or 3 times 10. Write the reference numbers in brackets, and write the second reference number as a multiple of the first.

$$(10 \times 3) \qquad 8 \times 27 =$$

Both the numbers in the example are below their reference numbers, so draw the circles below. Below the 8, which has the base reference number of 10, draw another circle.

$$(10 \times 3) \qquad 8 \ \times 27 =$$

How much are 8 and 27 below their reference numbers (remember the 3 represents 30)? Two and 3. Write 2 and 3 in the circles.

$$(10 \times 3) \qquad 8 \ \times 27 =$$
$$-②\ \ -③$$

Now multiply the 2 below the 8 by the multiplication factor, 3, in the brackets.

$$2 \times 3 = 6$$

Write 6 in the circle below the 2. Take the circled 6 diagonally from 27.

$$27 - 6 = 21$$

Multiply 21 by the base reference number, 10.

$$21 \times 10 = 210$$

To find the last part of the answer, multiply the two numbers in the top circles, 2 and 3, to get 6. Add 6 to our subtotal of 210 to get an answer of 216.

$$
\begin{array}{lll}
(10 \times 3) & 8 \times 27 = & 210 \\
& -②\ -③ & +6 \\
& -⑥ & \overline{216} \quad \textbf{Answer}
\end{array}
$$

Let's try another: 9×48.

Which reference numbers would we choose? I would choose 10 and 50. This is how we would write the problem:

$$
\begin{array}{ll}
(10 \times 5) & 9 \times 48 = \\
& \bigcirc \quad \bigcirc
\end{array}
$$

Both numbers are below the reference numbers, so we would draw the circles below. How much below? One and 2. Write 1 and 2 in the circles:

$$
\begin{array}{ll}
(10 \times 5) & 9 \times 48 = \\
& -① \ -②
\end{array}
$$

We now multiply the 1 below the 9 by the 5 in the brackets.

$$1 \times 5 = 5$$

We write 5 in a circle below the 1. This is how our problem looks now:

$$(10 \times 5) \qquad 9 \times 48 =$$
$$-① \quad -②$$
$$-⑤$$

We take 5 from 48:

$$48 - 5 = 43$$

Multiply 43 by the base reference number, 10 (write a zero after the 43 to get your answer).

$$43 \times 10 = 430$$

For the last step, we multiply the numbers in the original circles.

$$1 \times 2 = 2$$

Add 2 to our subtotal of 430.

$$430 + 2 = 432$$

The entire problem looks like this:

$$(10 \times 5) \qquad 9 \times 48 = 430$$
$$-① \quad -② \qquad +2$$
$$-⑤ \qquad \overline{432} \quad \textbf{Answer}$$

The calculation part of the problem is easy. The only difficulty you may have is remembering what to do next.

If the numbers are above the reference numbers, we do the calculation like this. We will take 13 × 42 as our example.

$$+\textcircled{12}$$
$$+\textcircled{3} \quad +\textcircled{2}$$
$$(10 \times 4) \qquad 13 \times 42 =$$

The base reference number is 10. The second reference number is 40, or 4 times 10. Try to keep the reference numbers both above or both below the numbers being multiplied. Both numbers in this example are above, so we draw the circles above. Thirteen has the base reference number of 10 so we draw two circles above the 13. How much above the reference numbers are 13 and 42? Three and 2. Write 3 and 2 in the circles. Multiply the 3 above 13 by the multiplication factor in the brackets, 4.

$3 \times 4 = 12$

Write 12 in the top circle above 13. Now add diagonally.

$42 + 12 = 54$

Fifty-four times our base number of 10 is 540. This is our subtotal. Now multiply the numbers in the first circles.

$3 \times 2 = 6$

Add 6 to 540 for our final answer of 546. This is how the finished problem looks:

$$+ \boxed{12}$$
$$+ \boxed{3} \quad + \boxed{2}$$
$$(10 \times 4) \qquad 13 \times 42 = 540$$
$$\underline{+\ 6}$$
$$546 \quad \textbf{Answer}$$

The base reference number does not have to be 10. To multiply 23×87 you would use 20 as your base reference number and 80 (20×4) as your second.

Let's try it:

$$(20 \times 4) \qquad 23 \times 87 =$$

Both numbers are above the reference numbers, 20 and 80, so we draw the circles above. How much above? Three and 7. Write 3 and 7 in the circles.

$$+ \boxed{3} \quad + \boxed{7}$$
$$(20 \times 4) \qquad 23 \times 87 =$$

We multiply the 3 above the 23 by the multiplication factor in the brackets, 4.

$$3 \times 4 = 12$$

Write 12 above the 3. Your work should look like this:

$$
\begin{array}{l}
+\text{⑫} \\
+\text{③} \quad +\text{⑦} \\
(20 \times 4) \qquad 23 \times 87 =
\end{array}
$$

Then add the 12 to the 87.

$87 + 12 = 99$

We multiply 99 by the base reference number, 20.

$99 \times 20 = 1980$

We multiply 99 by 2 and then by 10. Ninety-nine is 100 minus 1. Two times 100 minus 1 is 200 minus 2: $200 - 2 = 198$. Now multiply 198 by 10 to get our answer for $20 \times 99 = 1980$. Next, multiply the numbers in the original circles and add them to the subtotal.

$3 \times 7 = 21$

$1980 + 21 = 2001$

The finished problem looks like this:

$$
\begin{array}{l}
+\text{⑫} \\
+\text{③} \quad +\text{⑦} \\
(20 \times 4) \qquad 23 \times 87 = \quad 99 \\
\qquad\qquad\qquad\qquad\quad 1980 \\
\qquad\qquad\qquad\qquad\underline{+\ 21} \\
\qquad\qquad\qquad\qquad\quad 2001 \quad \textbf{Answer}
\end{array}
$$

Try these

Here are some more to try by yourself:

(a) $14 \times 61 =$

(b) $96 \times 389 =$

(c) $8 \times 136 =$

To calculate 8×136 you would use 10 and 140 (10×14) as reference numbers.

The answers are:

(a) 854   (b) 37 344   (c) 1088

Let's calculate (b) and (c) together:

$96 \times 389 =$

We use 100 and 400 as our reference numbers.

(100×4) $96 \times 389 =$
$-④$ $-⑪$

We multiply the 4 in the circle by the 4 in the brackets.

$4 \times 4 = 16$

We write 16 below the circled 4. Our work looks like this:

$$(100 \times 4) \qquad 96 \times 389 =$$
$$-④ \quad -⑪$$
$$-⑯$$

We now subtract 16 from 389 and get an answer of 373. We multiply the 373 by the base reference number, 100, to get an answer of 37300.

$$(100 \times 4) \qquad 96 \times 389 = 37300$$
$$-④ \quad -⑪$$
$$-⑯$$

We now multiply 4 by the 11 in the circle for an answer of 44. We add 44 to 37300 to get an answer of 37344. The completed problem looks like this:

$$(100 \times 4) \qquad 96 \times 389 = \quad 37300$$
$$-④ \quad -⑪ \qquad + 44$$
$$\overline{}$$
$$-⑯ \qquad\qquad 37344 \quad \textbf{Answer}$$

Let's try 8 × 136.

We use 10 and 140 (10 × 14) as our reference numbers.

$$(10 \times 14) \qquad 8 \times 136 =$$
$$-② \quad -④$$

We multiply the 2 in the circle by the 14 in the brackets.

$$2 \times 14 = 28$$

Write 28 below the circled 2. We subtract 28 from 136 (take 30 and add 2) to get 108. We multiply 108 by the base reference number, 10, to get 1080.

Our work looks like this so far:

$$(10 \times 14) \qquad 8 \times 136 = 1080$$
$$-② \quad -④$$
$$-㉘$$

We now multiply the numbers in the original circles.

$$2 \times 4 = 8$$

Add 8 to 1080 to get an answer of 1088. The completed problem looks like this:

$$(10 \times 14) \qquad 8 \times 136 = 1080$$
$$-② \quad -④ \qquad \quad +8$$
$$-㉘ \qquad \overline{1088} \quad \textbf{Answer}$$

Multiplying with fractions

To multiply 47×96, we could use reference numbers of (50×2) or $(100 \div 2)$. The latter would be easier because 100 then becomes our base reference number. It is easier to multiply by 100 than it is by 50. When writing the multiplication, I write the number first that has the base reference number.

$$(100 \div 2) \qquad 96 \times 47 = 4500$$
$$-④ \quad -③ \qquad +12$$
$$-② \qquad \overline{4512} \quad \textbf{Answer}$$

If we were multiplying 96 × 23, we could use 100 as our base reference number and 25 as our second reference number (100 ÷ 4 = 25). We would write the problem like this:

$$(100 \div 4) \qquad 96 \times 23 =$$
$$-④ \quad -②$$

Ninety-six is 4 below 100 and 23 is 2 below 25. We now divide the 4 in the circle by the 4 in the brackets. Four divided by 4 is 1. Write this below the circled 4.

$$(100 \div 4) \qquad 96 \times 23 =$$
$$-④ \quad -②$$
$$-①$$

Subtract 1 from 23 to get an answer of 22. Multiply the 22 by the base reference number, 100, for an answer of 2200.

Multiply the numbers in the original circles.

$$4 \times 2 = 8$$

Add this to 2200 to get an answer of 2208. The completed problem looks like this:

$$(100 \div 4) \qquad 96 \times 23 = \quad 2200$$
$$-④ \quad -② \qquad +8$$
$$-① \qquad \qquad \overline{2208} \quad \textbf{Answer}$$

What if we had multiplied 97 by 23? It would still work.
Let's try it:

$$(100 \div 4) \qquad 97 \times 23 =$$

$$-③ \quad -②$$

$$-¾$$

Three divided by four is three over four, or three quarters.
Take three quarters from 23. (Take one and give back
a quarter.)

$$23 - ¾ = 22¼$$

$$22¼ \times 100 = 2225 \text{ (25 is a quarter of 100)}$$

Multiply the numbers in the circles.

$$3 \times 2 = 6$$

$$2225 + 6 = 2231$$

$$(100 \div 4) \qquad 97 \times 23 = 2225$$

$$-③ \quad -② \qquad +6$$

$$-¾ \qquad \overline{2231} \qquad \textbf{Answer}$$

How about 88 × 343? We can use reference numbers of
100 and 350.

$$(100 \times 3½) \qquad 88 \times 343 =$$

$$-⑫ \quad -⑦ \qquad \underline{\qquad}$$

$$-㊷$$

To find the answer to 3½ × 12, you multiply 12 by 3, which is 36, and then add half of 12, which is 6, to get 42.

$343 - 42 = 301$

$301 \times 100 \text{ (base reference number)} = 30\,100$

$12 \times 7 = 84$

$30\,100 + 84 = 30\,184$

$(100 \times 3½)$ $88 \times 343 = 30\,100$
            $-⑫$ $-⑦$ $+\,84$
      $-㊷$ $\overline{30\,184}$ **Answer**

Why does this method work?

Here is my short explanation of why the method works. Let's take a look at 8 × 17. We could double the 8 to make 16, multiply it by 17, and then halve the answer to get the correct answer for the original problem. This is a hard way to go about it, but it will illustrate why the method using two reference numbers works. We will use a reference number of 20.

㉜ $16 \times 17 =$
     $-④$ $-③$

Subtract 4 from 17 and you get an answer of 13. Multiply the 13 by the reference number, 20, to get an answer of 260. Now multiply the numbers in the circles.

$4 \times 3 = 12$

Add 12 to the previous answer of 260 for a final answer of 272.

⑳ $16 \times 17 = 13$
      $-④\ -③$ 260
              $+ 22$
              272   **Answer**

But we multiplied by 16 instead of 8, so we have doubled the answer. Now 272 divided by 2 gives us our answer of 136.

$8 \times 17 = 136$

Now, we doubled our multiplier at the beginning and then halved the answer at the end. These two calculations cancel. We can leave out a considerable portion of the calculation to get the correct answer. Let's see how it works when we use the two reference number method.

(10×2) $8\ \times 17 = 130$
          $-②\ -③$ $+6$
       $-④$ 136   **Answer**

Notice that we subtracted 4 from 17 in the second calculation, the same as we did in the first. We got an answer of 13, which we multiplied by 10. In the first calculation we doubled the 13 before multiplying by 10, then we halved the answer at the end. The second time we ended by multiplying the original circled numbers, 2 and 3, to get an answer of 6, half the answer of 12 we got in the first calculation.

Tips

You can use any combination of reference numbers. The two general rules are:

- Make the base reference number an easy number to multiply by, for example, 10, 20, 50.

- The second reference number must be a multiple of the base reference number — for example, double the base reference number, 3 times, 10 times, 14 times and so on.

Play with the strategies. There is no limit to what you can do to make maths calculations easier. And each time you use these strategies you develop your mathematical skills.

Key points

- Make the base reference number an easy number to multiply by, for example, 10 or 100.

- Where possible, keep both reference numbers either above or below the numbers being multiplied. The final calculation will then be an addition instead of a subtraction.

Chapter 10
Using two reference numbers with factors

I was being interviewed on a radio talk program and I explained how students could master the multiplication tables in minutes. I explained how I multiply 7 times 8 and 96 times 97 using circles.

The interviewers said, we want to know how you would multiply 26 times 37. I said I would have to take them through a couple of intermediate explanations to explain it, but they insisted I explain how to multiply 26 × 37. I was not allowed to give my usual explanation: just do the calculation they told me. I had several options, using 20 or 30 as a reference, or even using two reference numbers, 20 and 40.

My choice wasn't good and my explanation fell flat which, I suspect, was what they wanted.

I have since realised how I should have answered. Almost any combination of numbers is easy to multiply if you use factors.

Let's use the example I was given.

An easy way to multiply 26 times 37 is to break up 26 into factors of 2×13. Then the problem can become $13 \times 2 \times 37$ or $13 \times (2 \times 37)$.

2×37 is 74. Our calculation is now 13×74. This is easily solved using 10 and 70 as reference numbers as 13 is close to an easy base number of 10.

$$(10 \times 7) \quad \overset{\text{\textcircled{3}} \quad \text{\textcircled{4}}}{13 \times 74} =$$

We multiply the 3 above 13 by the multiplication factor (in the brackets) 7 to get 21. We write 21 above the 3.

$$(10 \times 7) \quad \overset{\overset{\text{\textcircled{21}}}{\text{\textcircled{3}} \quad \text{\textcircled{4}}}}{13 \times 74} =$$

We add 21 to 74 to get 95. Multiply 95 by the base reference number, 10 to get 950. This is our subtotal.

Now multiply the numbers in the original circles and add.

$3 \times 4 = 12$

$950 + 12 = 962$ **Answer**

$$(10 \times 7) \quad \overset{\overset{\text{\textcircled{21}}}{\text{\textcircled{3}} \quad \text{\textcircled{4}}}}{13 \times 74} = \begin{array}{r} 950 \\ + 12 \\ \hline 962 \end{array} \textbf{ Answer}$$

This calculation can easily be done in your head.

Let's try another example, 38 × 73.

We can halve 38 to get 19 so the factors are 2 and 19. That would give us 73 × 2 × 19. Multiplying 73 by 2 we get 146 so the problem becomes 19 × 146. We can use 20 as a reference number and 140 (20 × 7) as our second.

$$(20 \times 7) \quad \overset{\textcircled{6}}{\underset{\textcircled{1}}{19 \times 146}} =$$
$$\textcircled{7}$$

We multiply 1 times 7 in the brackets to get the second number below 19. We subtract 7 from 146 to get 139.

To multiply 139 by 20 we double and then add a zero. 139 doubled is 278. (Twice 140 − 1 is 280 − 2.)

Multiply 278 by 10 to get 2780.

Multiply the numbers in the original circles.

$6 \times -1 = -6$
$2780 - 6 = 2774$ **Answer**

$$(20 \times 7) \quad \overset{\textcircled{6}}{\underset{\textcircled{1}}{19 \times 146}} = \quad 139$$
$$\begin{array}{r} 2780 \\ \textcircled{7} \quad -6 \\ \hline 2774 \quad \textbf{Answer} \end{array}$$

You could still halve 19 to get 9.5. The problem then becomes 9.5 × 292.

Now we can use 10 and 300 as reference numbers.

(10×30) $9.5 \times 292 = 2770$

$\textcircled{0.5}$ $\textcircled{8}$ $\underline{+4}$

$\textcircled{15}$ 2774 **Answer**

$30 \times 0.5 = 15$

$292 - 15 = 277$

$277 \times 10 = 2770$

Multiply the numbers in the circles: $8 \times 0.5 = 4$, or half of 8 is 4.

$2770 + 4 = 2774$

Using factors gives us a greater choice of methods for using our formula. Both of the methods are easier than direct multiplication.

Using the method with prime numbers

What do we do if we can't factorise the numbers? It doesn't matter. We can factorise any numbers like we did with 19 in the previous problem.

Let's try 31×73, two prime numbers.

Tip

A prime number is a number that can't be split into factors. The only factors of a prime number are the number itself and 1. Five is a prime number; its only factors are 5 and 1. So are 7, 11, 13, 17 and 19. Both 31 and 73 are also prime numbers.

We could double 31 to get 62 and then halve the answer. It would be easier to halve 31 to get 15.5 and multiply 73 by 2.

To multiply 73 by 2 you would break up 73 into 70 plus 3. If you see 70 as 7 times 10 you would multiply 7 by 2 to get 14 and then by 10 to get 140. Two times 3 is 6. Add 6 to 140 for the answer, 146. The problem now becomes 15.5 by 146.

$$\begin{array}{cc} \quad \overset{\textstyle (5.5)}{} \quad \overset{\textstyle (6)}{} \\ (10 \times 14) \quad 15.5 \times 146 = \end{array}$$

Multiply 5.5 (above 15.5) by 14 (multiply 5.5 by 2 × 7)

5.5 × 2 is 11, multiplied by 7 is 77.

$$\begin{array}{cc} \quad \overset{\textstyle (77)}{\overset{\textstyle (5.5)}{}} \quad \overset{\textstyle (6)}{} \\ (10 \times 14) \quad 15.5 \times 146 = \end{array}$$

Add 77 to 146 to get 223. (Add 80, subtract 3 or add 100 in your head, then subtract 20 and then subtract 3. With practice you will find this easy.) Multiply by 10 to get 2230.

Now multiply the numbers in the original circles:

$6 \times 5.5 = 33$

$2230 + 33 = 2263$ **Answer**

An alternative method for multiplying 31 by 73 would be to multiply 73 by 30 and then add 73 for the answer.

To multiply 96 by 194 you can factorise 194 as 2×97.

96×97 is 9312. Then 9312 doubled is 18 624.

Or you can use two reference numbers, 100 and 200.

(100×2) $96 \times 194 =$
                      ④   ⑥
                      ⑧

Subtract 8 from 194 to get 186. Multiply 186 by the base reference number, 100, to get 18,600. Then multiply the numbers in circles:

$4 \times 6 = 24$

$18\,600 + 24 = 18\,624$ **Answer**

The full calculation looks like this:

(100×2) $96 \times 194 = 18\,600$
                      ④   ⑥ $+ 24$
                      ⑧ $\overline{18\,624}$ **Answer**

Both of these methods are easier than standard multiplication.

 Try these

Here are some problems to try for yourself:

(a) 24 × 37 =

(b) 71 × 26 =

(c) 33 × 72 =

(d) 28 × 54 =

Here are the answers:

(a) 888   (b) 1846   (c) 2376   (d) 1512

With imagination and a little experimentation you can use this method to multiply any numbers.

Key points

- Welcome challenges: every time someone challenged my methods it resulted in improvements.

- Prime numbers can always be halved.

Chapter 11
Learning beyond the tables

When you know your multiplication tables up to the 5 times table you can multiply pretty much anything. If you know your 2, 3, 4 and 5 times table up to 10 times each number, then you also know part of your 6, 7, 8 and 9 times tables. For the 6 times table, you also know 2 times 6, 3 times 6, 4 times 6 and 5 times 6. The same applies to the 7, 8 and 9 times tables.

As you learn your higher tables, mental calculations become much easier. It is easy to learn your 11 times table and the 12 times table is not difficult. Twelve is 10 plus 2, so to multiply by 12 you can multiply the number by 10 and then add 2 times the number. Six times 12 is 6 times 10, plus 6 times 2. Six times 10 is 60, and 6 times 2 is 12. It is easy to add the answers for 6 times 12 to get 72. You could also use factors and multiply by 6 and then by 2, or multiply by 2 and then by 6.

Learn the 13, 14 and 15 times tables and more

It is usual in schools to teach tables up to the 12 times table. It is not hard to learn the 13 times table if you know the 12 times table. If you know 12 threes are 36, 13 threes are just three more, so just add another 3 to get 39.

Twelve threes plus one more 3 makes 13 threes. If you know 12 fours are 48, just add another 4 to get 52.

When you know your 13 times table, you can easily learn the 14 times table. You have two choices. You can factorise 14 to 2 times 7. Then you just double the number and multiply by 7, or multiply by 7 and double the answer. So, 6 times 14 would be 6 times 7 (42), which when doubled is 84. That was easy. You could also have doubled the 6 first to get 12, and multiply 12 by 7 to get 84 again. If you know your 12 times table, that was easy as well. Another alternative is to say 6 times 10 is 60, plus 6 times 4 is 24. Add 60 and 24 for the answer, 84.

Calculate these often and you will learn your tables without even trying. Then you will find you can multiply and divide directly by 13, 14 and 15. This will give you an advantage over the other kids.

I used to give out tables sheets to my private students to do for homework. The tables sheets went up to 19 times 19. I would correct their work each week. I never used an answer sheet—I had to calculate the answers by myself to see if their answers were correct. After correcting 14 times 14 was 196 a couple of dozen times I found I remembered the answer. I had memorised 14 times 14. The same went for 15 times 15, 16 times 16, 17 times 17 and so on. I learnt my tables the same way my students did. It was a much more pleasant method than chanting the tables like I did at primary school.

This is the method I advocate for students to learn all of their tables. It is no harder to learn 7 times 8 is 56 than to remember 14 times 14 is 196. And, for a while when I was correcting the tables sheets, I remembered the answers but still did a lightning calculation just to make sure.

Tip

Some of my students remembered the answers, but they somehow thought it was immoral to write the answer from memory, as if they were lazy or avoiding work. I found the lazy students learnt their tables faster than those who felt they had to work at their tables.

Three methods for multiplying by numbers in the teens

There are three basic methods I recommend for multiplying numbers in the teens. We already have an easy method for multiplying two numbers in the teens. To multiply 14 times 14 we use circles to find the answer 196. But, to multiply a single digit number by a number in the teens we have a choice of three methods.

1   Use the above and below circles method.

2   Multiply the number by ten and then by the units digit, then add.

3   Use factors if the teens number can be factorised.

Let's use all three methods to multiply 7×14.

First method

$$\overset{\text{\textcircled{4}}}{\underset{\text{\textcircled{3}}}{\text{\textcircled{10}} \quad 7 \times 14 =}}$$

$14 - 3 = 11$

11×10 (reference number) $= 110$

$3 \times 4 = 12$

$110 - 12 = 98$

(Subtract 10, then 2, to get the answer.)

Second method

$7 \times 10 = 70$

$7 \times 4 = 28$

$70 + 28 = 98$

Third method

The factors for 14 are 2×7

$7 \times 14 = 7 \times 7 \times 2$

$7 \times 7 = 49$

$49 \times 2 = 98$

The easy way to calculate 2 times 49.

($49 = 50 - 1$. Then 2 times $50 - 1$ is $100 - 2$)

All three methods are fairly easy. You don't have to use the same method all the time; you can choose which method is easiest for the numbers you have to multiply. You can decide which method you prefer.

For instance, to multiply 7×15, you already have the answer to fourteen sevens so you can just add another 7 to make it fifteen sevens: $98 + 7 = 105$.

That used a different method again.

To multiply by 17, 18 and 19 you might find it easier to use the circle method of multiplication.

Learn your tables as high as you like

You can take your tables as high as you like. Why stop at the 15 times table, or even the 20 times table? Multiplying by 16 is easy if you know the 8 times table. You simply factorise 16 to 2 times 8. To multiply 4 times 16 you multiply 4 by 2 times 8, or you multiply 4 by 8 times 2 for the same answer.

To multiply 4 by 2 times 8 you would multiply 4 by 2 to get 8, and then multiply by 8 to get 64. If you didn't know the answer to 8 times 8 you could simply use the circles to calculate the answer.

To multiply 4 by 8 times 2 you would multiply 4 by 8 to get 32 and double the answer to get 64. Two thirties are 60 plus 2 times 2 is 4. Remember, you can multiply the factors in any order you like.

Here are some suggestions for multiplying by the numbers in the teens when you are multiplying a single digit number.

Multiply by 13

Multiply by 12 and add the number.

Multiply by 10 and add three times the number.

Let's look at 4×13.

Four times 12 is 48, plus another 4 equals 52.

You could also multiply by 10 and add 3 times the number. To multiply 4×13 you would multiply 4 by 10 (40) and add 3 times 4. Then 40 plus 12 will give you 52.

To multiply 7 times 13, if you aren't sure of 7×12 you might prefer to multiply 7 by 10 (70) and then add 3 times 7 equals 21. Then 70 plus 21 equals 91.

Multiplication by 13 is easy.

Multiply by 14

Fourteen is 2 times 7 so you can multiply by 7 and double your answer, or multiply by 10 and add 4 times the number.

For example, 6×14 can be calculated by factorising 14 as 2×7.

Six times 2 is 12, times 7 is 84.

Or you can multiply $6 \times 7 = 42$, and then double your answer for 84.

If you don't like either way of using factors, you can calculate $6 \times 10 = 60$, plus $6 \times 4 = 24$.

$60 + 24 = 84$

Play with the methods until you don't need them; you will remember the answers.

Multiply by 15

Choose one of these methods:

1   Multiply by 10 and add half the answer.

2   Multiply by 10 and add 5 times the number.

3   Multiply by 5 and then the answer by 3 ($3 \times 5 = 15$).

Here's an example:

$8 \times 15 =$

Using method 1 we multiply 8 by 10 and add half the answer.

$8 \times 10 = 80$

Half of 80 is 40.

$80 + 40 = 120$

Using factors we can multiply $8 \times 5 \times 3$.

$8 \times 5 = 40$

$40 \times 3 = 120$

Multiply by 16
Multiply by 8 and double the answer.

Multiply by 17
Multiply by 10 and add 7 times the number.

Or you can use above and below circles.

Multiply by 18
Multiply by 9×2.

Or you can use above and below circles.

Multiply by 19
Multiply by 10 and add 9 times the number.

Or multiply by 20 and subtract the number.

Or you can use above and below circles.

Multiply by 20

Use factors. Double the number and add a zero. (Multiply by 2 and the answer by 10.)

 Try these

Do the following calculations in your head.

(a) $4 \times 13 =$

(b) $6 \times 15 =$

(c) $7 \times 16 =$

(d) $8 \times 13 =$

(e) $9 \times 13 =$

The answers are:

(a) 52   (b) 90   (c) 112   (d) 104   (e) 117

You don't have to formally learn your higher tables. Just practise the calculations so you can solve them quickly and people won't know if you have them memorised or not. You will find that you will remember them and you can call the answers immediately and, maybe, make a lightning calculation to be sure you have the right answer.

Tip

If you are just checking you haven't made a mistake you could always cast out nines if you don't feel like doing the calculation.

Now you have a whole lot of choices.

Key points

- Learning the tables saves work.
- Use your brain instead of a calculator.

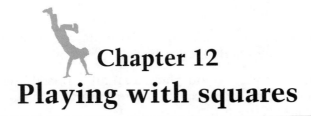

Chapter 12
Playing with squares

You will find there are many calculations where you need to square numbers. In engineering, electronics and electrical studies, calculating areas of circles, astronomy, calculating right-angled triangles, light measurement and more, you often need to find the square of numbers.

Tip

Squaring a number means multiplying a number by itself.

For instance, 5 squared means 5×5. It is written as 5^2. The small 2 written after the 5 means there are two fives multiplied together. If you see a small 3 written after the 5 it means there are three fives to be multiplied. This is a common mathematical procedure and one that everyone should know. Here are a few examples.

$$5^3 = 5 \times 5 \times 5$$

$$4^5 = 4 \times 4 \times 4 \times 4 \times 4$$

$$7^4 = 7 \times 7 \times 7 \times 7$$

So 5^3 means three fives multiplied together, 4^5 means five fours multiplied together and 7^4 means four sevens multiplied together. Using the small number alongside the number to be multiplied is just an easier way to say it and to write it.

Let us look at an example. To square 4 you would multiply 4 by itself: $4 \times 4 = 16$

To write 4 squared you place a small 2 raised after the 4 to show that you mean two fours multiplied together. You would write four squared as 4^2.

So, $4^2 = 16$.

Twelve squared is $12 \times 12 = 144$.

We write this as $12^2 = 144$.

 Try these

Try these for yourself:

(a) 3^2

(b) 6^2

(c) 5^2

(d) 8^2

Those were easy.

Here are the answers:

(a) 9　　(b) 36　　(c) 25　　(d) 64

Using factors

To square 30 you would multiply 30 by 30. Is this difficult? No because you would use factors. Thirty is 3×10 so you would multiply $(3 \times 10) \times (3 \times 10)$. The easy way to do this would be to multiply $3 \times 3 \times 10 \times 10$. Three by 3 is 9 and 10 times 10 is 100, so the answer is 900.

How would you calculate 40^2? Forty squared means 40×40. You would simply multiply the fours and add two zeros to the answer.

$4 \times 4 = 16$

Adding two noughts, or zeros, gives us 1600.

How would you calculate 50^2? See if you can calculate the answer in your head.

The easy way to calculate the answer is to multiply five times five and add two zeros to the answer.

$5 \times 5 = 25$

Adding two noughts or zeros we get an answer of 2500.

That was easy, wasn't it?

Try these

Now do these by yourself.

(a) 20^2

(d) 80^2

(b) 60^2

(e) 110^2

(c) 90^2

(f) 120^2

Did you find those easy?

The answers are:

(a) 400	(b) 900	(c) 8100
(d) 6400	(e) 12100	(f) 14400

Squaring numbers ending in 5

Squaring numbers ending in 5 is just as easy as squaring numbers ending in zero. The method for numbers ending in 5 uses the same formula we have used for general multiplication. It is a fun way to play with the strategies you have already learnt.

To square a number ending in 5, separate the final 5 from the digit or digits that come before it. Add 1 to the number in front of the 5, then multiply the two numbers together. Write 25 (which is 5 squared) at the end of the answer and the calculation is complete.

Example one

Let's try 35^2 as an example.

Separate the 5 from the digits in front—in this case there is only a 3. Add 1 to the 3 to get 4. Multiply these numbers together:

$3 \times 4 = 12$

Write 25 after the 12, which gives us an answer of 1225.

Example two

Here is another example—75^2, or 75 squared.

Separate the 7 from the 5. Add 1 to the 7 to get 8. Eight times 7 is 56. This is the first part of our answer. Write 25 at the end of our answer and we get 5625.

We can combine methods to get even more impressive answers. Let's try 135^2.

Separate the 13 from the 5. Add 1 to 13 to give 14. Thirteen times 14 is 182 (we use the method taught in chapter 3). Write 25 at the end of 182, which gives us an answer of 18225. This can easily be calculated in your head.

Example three

Here is another example—965^2.

Ninety-six plus 1 is 97. Multiply 96 by 97, which gives us 9312. Write 25 at the end, which gives us an answer of 931225.

That is impressive, isn't it?

Try these

Try these for yourself:

(a) 15^2 (e) 95^2

(b) 45^2 (f) 115^2

(c) 25^2 (g) 145^2

(d) 65^2 (h) 955^2

The answers are:

(a) 225   (b) 2025   (c) 625   (d) 4225
(e) 9025   (f) 13 225   (g) 21 025   (h) 912 025

If you used pencil and paper to calculate the answers, do them again in your head. You will find it quite easy.

Why it works

Why does this work? Let's take the example of 25^2 and use our method for multiplying with circles. We will use 20 as a reference number.

$$\textcircled{\scriptsize 5} \quad \textcircled{\scriptsize 5}$$
$$\textcircled{\scriptsize 20} \quad 25 \times 25 =$$

We add crossways: $25 + 5 = 30$

We multiply the answer, 30, by the reference number, 20.

$20 \times 30 = 600$

Now we multiply the numbers in the circles.

$5 \times 5 = 25$
$600 + 25 = 625$ **Answer**

Here is the calculation in full.

$$\begin{array}{r} \text{⑤} \quad \text{⑤} \\ \text{⑳} \quad 25 \times 25 = \quad 600 \\ 25 \\ \hline 625 \end{array}$$

Can you see that the short cut actually uses the method we have been using?

Multiplying two numbers where tens digits are the same and units digits add to 10

If the units digits add to ten and the tens digit is the same, then we can use the general rule for squaring numbers ending in 5 but with a slight difference. We add 1 to the tens digit and multiply the two numbers. Multiply the answer by 100 (add two zeroes). Then multiply the units digits and add to your previous answer.

Example one

Let's try 24 × 26 as an example.

Add 1 to the tens digit and multiply.

$2 + 1 = 3$

$2 \times 3 = 6$

This is the hundreds digit of the answer so the first step in the answer is really 600.

Now we multiply the units digits and add the answer.

$4 \times 6 = 24$

$600 + 24 = 624$ **Answer**

23 × 27 would be 621 and 22 × 28 would be 616.

Example two

Now let's try 43 × 47 =

The units digits add to 10: $3 + 7 = 10$.

We add 1 to the tens digit: $4 + 1 = 5$.

Now multiply $4 \times 5 = 20$.

Now multiply by 100: $20 \times 100 = 2000$.

Then multiply the units digits and add: $3 \times 7 = 21$

$2000 + 21 = 2021$ **Answer**

What would the answer be to 31 × 39? Here's how we work it out:

$3 + 1 = 4$

$3 \times 4 = 12$

$12 \times 100 = 1200$

$1 \times 9 = 9$

$1200 + 9 = 1209$ **Answer**

 Try these

Try these for yourself:

(a) 24 × 26 =

(b) 62 × 68 =

(c) 37 × 33 =

(d) 42 × 48 =

(e) 64 × 66 =

(f) 88 × 82 =

The answers are:

(a) 624	(b) 4216	(c) 1221
(d) 2016	(e) 4224	(f) 7216

Multiplying numbers near a square

Here is an interesting property of squares. If you go one above a number and one below and multiply them the answer is one less than the number squared.

Let's try it with the number 4.

Four squared (4×4) equals 16.

If you go one above 4 (5) and one below 4 (3) and multiply 5×3 you get an answer of 15, one less than 4 squared.

Fourteen squared is 196. Multiplying one above and one below, $13 \times 15 = 195$.

Now let's look at 5 squared is 25.

If you go one above 5 (6) and one below 5 (4) and multiply 6×4 you get an answer of 24, one less than 5 squared equals 25.

If you go two above 5 (7) and two below 5 (3) and multiply 7×3 you get an answer of 21, which is 4 (or 2 squared) less than 5 squared equals 25.

If you go three above 5 (8) and three below 5 (2) and multiply 8×2 you get an answer of 16, which is 9 (or 3 squared) less than 5 squared equals 25.

This makes it easy to multiply numbers that are an equal amount above and below a number that is easy to square.

Let's say we want to multiply 39 by 41. The numbers are one above and one below 40, which is easy to square.

Forty squared is 1600 $(4 \times 4 \times 10 \times 10)$.

One squared is 1 so we subtract 1 from 4 squared $(1600 - 1)$ to get 1599.

Let's try another: $28 \times 32 =$

The numbers are two above and two below 30.

Thirty squared is 900.

Two squared is 4.

900 minus 4 is 896.

Let's try one more.

$65 \times 75 =$

The numbers are five above and below 70.

Seventy squared is 4900.

Five squared is 25.

$4900 - 25 = 4875$ **Answer**

The next calculation is impressive.

$137 \times 143 =$

The numbers are 3 above and below 140.

It is easy to find the square of 140. Just multiply 14 by 14 using the short cut for multiplying numbers in the teens and add 2 zeros to the answer.

$14 \times 14 = 196$

Adding two zeros we get 19600.

Three squared is 9.

Subtract 9 from 19600 to get 19591. (Remember, to subtract 9 we take away 10 and add 1.)

Try these

Now try these for yourself:

(a) 49 × 51 =

(b) 38 × 42 =

(c) 47 × 53 =

(d) 136 × 144 =

The answers are:

(a) 2499   (b) 1596   (c) 2491   (d) 19 584

Are you impressed by what you are doing? You will impress your family, friends and classmates by doing those calculations in your head.

Short cut for squaring numbers near 50

Our method for multiplying numbers near 50 uses the same formula as for general multiplication but, again, there is an easy short cut.

Example one: 46²

46^2 means 46 × 46.

50 × 50 = 2500. We take 50 and 2500 as our reference points.

46 is below 50 so we draw a circle below.

$$\overset{\textstyle 50}{\underset{-\,\textstyle 4}{\quad}} \quad 46^2 =$$

46 is 4 below 50, so we write a 4 in the circle. It is a minus number as 46 is 50 minus 4.

We take 4 from the number of hundreds in 2500.

$25 - 4 = 21$. That is the number of hundreds in the answer. Our subtotal is 2100.

To get the rest of the answer, we square the number in the circle.

$4^2 = 16$.

$2100 + 16 = 2116$ **Answer**

Example two: 56^2

56 is above 50 so draw the circle above.

$$\overset{+\,\textstyle 6}{\underset{\textstyle 50}{\quad}} \quad 56 =$$

We add 6 to the number of hundreds in 2500: $25 + 6 = 31$.

Our subtotal is 3100.

$6^2 = 36$.

$3100 + 36 = 3136$ **Answer**

Example three: 62²

$\overset{\text{⑫}}{\underset{\text{㊿}}{}}$ 62 =

25 + 12 = 37. Our subtotal is 3700.

$12^2 = 144$

$3700 + 144 = 3844$ **Answer**

 Try these

Practise with these:

(a) 57^2

(b) 51^2

(c) 48^2

(d) 39^2

(e) 45^2

The answers are:

(a) 3249 (b) 2601 (c) 2304 (d) 1521 (e) 2025

With a little practice, you should be able to call the answer without a pause.

Tip

If you can square numbers in your head, you will be known as a prodigy.

Units digits add to 10 and tens digits differ by 1

If the units digits add to 10 and the tens digits differ by 1, then you will find the numbers to be multiplied are an equal amount above and below the higher tens number.

For example, let's take the calculation 23×37. The tens digits differ by 1 and the units digits add to 10 $(3 + 7 = 10)$. The numbers are 7 above and 7 below 30. It is easy to square 30: 30^2 $(30 \times 30) = 900$.

The numbers are 7 above and 7 below 30. We square the 7 to get 49 $(50 - 1)$.

Subtract 49 from 900.

$900 - 49 = 851$

(Subtract 50 and add 1.)

Try these

(a) 27 × 33 =

(b) 16 × 24 =

(c) 34 × 46 =

(d) 58 × 62 =

The answers are:

(a) 891   (b) 384   (c) 1564   (d) 3596

Key points

- You will find as you progress that you need to square numbers in many fields of study.

- Practise doing all of the exercises in your head.

- Using these methods will make squaring numbers easy and you will impress your friends.

Chapter 13
Direct multiplication

Everywhere I teach my methods around the world, students and teachers ask me, 'How would you multiply these numbers?' They are always looking for the exceptions that won't work with my methods. I like being asked these questions because it means my students are thinking seriously about my strategies and they are thinking mathematically. The kinds of questions they ask usually show that they understand the concepts.

From time to time I am given numbers that don't lend themselves to the methods with reference number and circles. When this happens I tell people I use direct multiplication.

Multiplication with a difference

If I were asked to multiply 7 times 13, I wouldn't use the method with the circles because I don't believe it is the easiest way to solve this problem. Instead, I would use direct multiplication. I would multiply 7 times 10 (from the 13) and then add 7 times 3.

$7 \times 10 = 70$

$7 \times 3 = 21$

$70 + 21 = 91$ **Answer**

How about 8 times 37?

Eight times 30 is 240 ($8 \times 3 \times 10 = 240$). Eight times 7 is 56. Then 240 plus 56 equals 296. You would say, 'Two hundred and forty plus fifty is two hundred and ninety, plus six is two hundred and ninety-six'.

It is easy to multiply any two-digit number by a one-digit number. This will enable you to use 60, 70 and 80 as reference numbers. This means there is no gap in the numbers up to 100 that are easy to multiply using the reference number and circles.

Let's try an example: 76×76.

We can either use 70 or 80 as reference numbers. Let's try with each. First we will use 70.

$$
\begin{array}{c}
\quad +\!\!\;⑥ \quad +\!\!\;⑥ \\
⑦⓪ \quad\quad 76 \times 76 =
\end{array}
$$

$76 + 6 = 82$

We multiply 82 by the reference number, 70. To do this we multiply by 7 and then by 10 because 70 is 7×10.

$7 \times 80 = 560$

$7 \times 2 = 14$

$560 + 14 = 574$

$574 \times 10 = 5740$

Our working so far looks like this:

$$
\begin{array}{c}
\quad +\!\!\;⑥ \quad +\!\!\;⑥ \\
⑦⓪ \quad\quad 76 \times 76 = 5740
\end{array}
$$

Now we multiply the numbers in the circles and add that answer to the subtotal.

$6 \times 6 = 36$

$5740 + 36 = 5776$

The completed calculation looks like this:

$$+ ⑥ \quad + ⑥$$
$$⑦⓪ \quad 76 \times 76 = 5740$$
$$+ 36$$
$$\overline{5776} \quad \textbf{Answer}$$

Let's try it again with 80 as a reference number.

$$⑧⓪ \quad 76 \times 76 =$$
$$- ④ \quad - ④$$

$76 - 4 = 72$

We multiply 72 by the reference number.

$8 \times 70 = 560$

$8 \times 2 = 16$

$560 + 16 = 576$

$576 \times 10 = 5760$

Our working so far is shown here.

$$⑧⓪ \quad 76 \times 76 = 5760$$
$$- ④ \quad - ④$$

Multiply the numbers in the circles and add that answer to the subtotal.

$4 \times 4 = 16$

$5760 + 16 = 5776$

The completed calculation looks like this:

$$⑧⓪ \qquad 76 \times 76 = 5760$$
$$-④ \quad -④ \quad + 16$$
$$\overline{5776} \qquad \textbf{Answer}$$

Direct multiplication by a two-digit number

Here is how I multiply by two-digit numbers using direct multiplication. Let's take a look at an example: 143×21.

We set the problem out like this:

$$143$$
$$\times 21$$

We do the multiplication in one step. We multiply each of the digits in turn by the tens digit of the multiplier, then the units digit. Let me show you what I mean.

We begin by multiplying the 3 in 143 by 20, then by 1. To multiply by 20, we multiply by 2 and then by 10.

$2 \times 3 = 6$

$6 \times 10 = 60$

$1 \times 3 = 3$

$60 + 3 = 63$

Write 3 as the units digit of the answer and carry the 6. We move to the next digit, 4, and do the same.

$2 \times 4 = 8$

$8 \times 10 = 80$

$1 \times 4 = 4$

$80 + 4 = 84$

We add the 6 that was carried and get 90. Write 0 as the next digit of the answer and carry the 9.

The problem looks like this so far:

$$\begin{array}{r} 143 \\ \times\, 21 \\ \hline {}^9 0\, {}^6 3 \end{array}$$

We now move on to the final digit to multiply, 1.

$20 \times 1 = 20$

$1 \times 1 = 1$

$20 + 1 = 21$

We now add the 9 that we carried.

$21 + 9 = 30$

The answer is 3003.

The full calculation is shown here.

$$\begin{array}{r} 143 \\ \times\,21 \\ \hline 30^90^63 \end{array}$$

Let's check our answer by casting out the nines.

$$143 \times 21 = 3003$$
$$8 \qquad 3 \qquad\; 6$$

$$8 \times 3 = 24$$
$$2 + 4 = 6$$

Our answer is correct—it is the same as the substitute number for 3003.

Let's try a longer calculation: 4235×42. We set out the problem like this:

$$\begin{array}{r} 4235 \\ \times\,42 \\ \hline \end{array}$$

And calculate as before.

$$4 \times 5 = 20$$
$$20 \times 10 = 200$$
$$2 \times 5 = 10$$
$$200 + 10 = 210$$

Write 0 and carry 21.

$$
\begin{array}{r}
4235 \\
\times\,42 \\
\hline
^{21}0
\end{array}
$$

We move to the next digit to multiply, 3.

$4 \times 3 = 12$

$12 \times 10 = 120$

$2 \times 3 = 6$

$120 + 6 = 126$

We add the 21 that we carried.

$126 + 21 = 147$

Write 7 as the next digit of the answer and carry 14.

$$
\begin{array}{r}
4235 \\
\times\,42 \\
\hline
^{14}7^{21}0
\end{array}
$$

We move to the next digit to multiply, 2.

$4 \times 2 = 8$

$8 \times 10 = 80$

$2 \times 2 = 4$

$80 + 4 = 84$

(Doing this in your head you would just say to yourself, 'Eighty…four'.)

Add the 14 carried.

$$84 + 14 = 98$$

Write 8 and carry 9.

$$4235$$
$$\times 42$$
$$\overline{{}^9 8 {}^{14} 7 {}^{21} 0}$$

Now for the final digit.

$$4 \times 4 = 16$$
$$16 \times 10 = 160$$
$$2 \times 4 = 8$$
$$160 + 8 = 168$$

Add the 9 that we carried.

$$168 + 9 = 177$$

Write down 177 as the last part of the answer.

$$4235$$
$$\times 42$$
$$\overline{177 {}^9 8 {}^{14} 7 {}^{21} 0}$$

The answer is 177 870.

Tip

You can multiply numbers of any length by a two-digit number using the direct multiplication method.

Try these

Following are some problems for you to try:

(a) $235 \times 41 =$

(b) $621 \times 37 =$

(c) $63 \times 46 =$

How did you go?

The answers are:

(a) 9635   (b) 22 977   (c) 2898

Did you get them all right? Don't worry if you made a mistake. This is a new procedure and it is easy to make mistakes in the beginning. With practice you will find this method much easier than the traditional method.

Let's do the first problem together.

$$235$$
$$\times\,41$$

First we multiply 5 by 40 and then by 1.

$4 \times 5 = 20$

$20 \times 10 = 200$

$1 \times 5 = 5$

$200 + 5 = 205$

Write 5 and carry 20.

$$235$$
$$\times\,41$$
$$\overline{^{20}5}$$

Now we multiply 3 by 40 and then by 1.

$4 \times 3 = 12$

$12 \times 10 = 120$

$1 \times 3 = 3$

$120 + 3 = 123$

Add the 20 that we carried.

$123 + 20 = 143$

Write 3 and carry 14.

$$
\begin{array}{r}
235 \\
\times\,41 \\
\hline
{}^{14}3{}^{20}5
\end{array}
$$

Now we multiply 2 by 41.

$$40 \times 2 = 80$$

$$1 \times 2 = 2$$

$$80 + 2 = 82$$

Add the 14 that we carried.

$$82 + 14 = 96$$

Write 96 as the final part of the answer.

$$
\begin{array}{r}
235 \\
\times\,41 \\
\hline
96{}^{14}3{}^{20}5
\end{array}
$$

If you made a mistake with either of the other two problems, go over them again and correct them. If you still can't find where you went wrong, then do the calculation in full using the traditional method. That should show you where you made the mistake.

This is a very useful way to multiply any numbers by a single-digit number and by a two-digit number.

Practise some more until you can do them without hesitation. With practice, you won't ever want to go back to the old way of multiplication.

There are occasions, however, when even using reference numbers or direct multiplication may prove difficult. For example, a young student in a Singapore school asked me, 'How would you multiply 16 359 482 by 5 718 295?'

I said, 'Step one, I would take out my calculator'.

Key points

- Try to find the easiest method to solve the problem.

- Direct multiplication can be carried out mentally.

- The answer can be calculated in one line.

Part II
Having fun with numbers

In this section we will play with numbers and explore their properties. We will look at solving puzzles and problems and what you can do if you have no idea where to start. Solving problems and puzzles is a learned skill.

Chapter 14
Solving problems and puzzles

There are many ways you can help your children develop their thinking skills. In particular, they can work with puzzles and problems to develop skills in problem-solving. In this chapter I explain how to solve maths problems and how to develop problem-solving skills. I have also included several puzzles for you to put your skills into practice.

Helping your child develop intelligence

One of the best ways to improve your children's thinking skills is to provide them with puzzles and problems to work through. Subscribe to a puzzle magazine that has problems that provide a challenge, but are not too hard. Puzzle books with detailed explanations of the answers can be very useful in teaching children and adults how to solve puzzles. Ask friends or family who are good at solving problems to explain their thinking methods. Sometimes people have trouble explaining their thinking strategies. Some will just say, 'I know. That's all. I don't have a method'. They do have a method: they are just not able to explain it. By making them think about their methods, you might be doing them a favour as well.

Here is an example of a fairly easy maths problem.

A bat and three balls cost $14.50.

Five balls cost $12.50.

(a) How much would two bats and four balls cost?

(b) How much does a bat cost?

(c) How much does a ball cost?

Many children will look at this problem with dismay and say they can't solve it. Here are some questions to ask to help the child find the answer:

- Which question can be answered with the information you have been given?

- What have you been told?

- What can you work out from what you have been told?

- Will it help to make a chart?

- What calculation can you make?

- When you have made your first calculation, can you then make a second calculation?

Find a place to start

You don't have to answer the first question first. It is important to be able to find where to start on a logic problem.

What is the first calculation you can make? The first statement, a bat and three balls cost $14.50, doesn't get us far because we don't know how much the bats or the balls cost by themselves. If we can't use the first statement, we don't give up. We go to the next statement—five balls cost $12.50.

Can we work out the cost of one ball from this statement? Yes, we can. We divide $12.50 by five to find the cost of one ball: $12.50 ÷ 5 = $2.50. We have answered the last question first.

Tip

Explain to your child that the questions can be answered in any order.

How do we calculate the cost of a bat? We go to the first statement. A bat and three balls cost $14.50. Now we know the cost of a ball, we can work out the cost of the bat. A bat and three balls cost $14.50, so a bat costs $14.50 less the cost of three balls. How much do three balls cost? Three times $2.50: 3 × $2.50 = $7.50. So, the cost of a bat is $14.50 less the cost of three balls, $7.50: $14.50 − $7.50 = $7.00. We have answered the second question.

Now we have answered the second and third questions, it is easy to answer the first.

How much would two bats and four balls cost? Two bats cost 2 times $7.00, which is $14.00. Four balls cost 4 times $2.50, which is $10.00. Then $14 plus $10.00 is $24.00, the answer to the first problem.

Students need to learn to reason this way to succeed in maths and increase their job opportunities when they leave school.

Developing problem-solving skills

Another way to improve your children's intelligence is by helping them develop problem-solving skills. Following are some rules or suggestions for developing problem-solving skills, and some puzzles for you and your children to try. The solutions to these puzzles can be found in appendix A.

Here are my rules for solving problems.

Work on the assumption you can solve the problem

If children believe they can't solve a problem, it's very likely they won't. They won't put in the same effort. They won't look for somewhere to start. Children who believe they can solve puzzles will look for somewhere to begin.

Tip

Give children plenty of encouragement. Applaud their efforts. Never tell them they are stupid if they fail. Tell them they didn't succeed because they didn't know how, not because they weren't intelligent.

Simplify numbers

There are many problems that appear difficult because the numbers are complex, but if you tried the same problem with simple numbers you could solve it immediately. For example, if you had to find 4 per cent of $47.36. What if it were $1.00 or $100.00 instead? How do you find 4 per cent of $47.36? Think about what 4 per cent of 100 is.

That's easy, 4. What did you do to find it? Do the same with $47.36.

You can also use this method to check that a strategy you are using to solve a problem is valid. Will your method work for all values? Does it work in this case where the answer is obvious? If it does, then you can proceed with the more difficult problem.

Work backwards

How do you find the value of a $16.45 item before 7 per cent tax was added? Most people would have difficulty with this. Many would find 7 per cent of $16.45, and they would be wrong. It was 7 per cent of the original price, which we don't know. So, how can we find it?

If you have no idea, do a simple problem backwards (this combines the second and third rules). Let's say we have an item that costs one dollar, plus 7 per cent tax. What is the total cost? One dollar and seven cents. How did we calculate that? By finding 7 per cent of a dollar, or 7 parts for each hundred, and adding it. Now we have a total cost of $1.07 and we want to find the original price before we added 7 per cent. We know the answer is one dollar, but what do we do to $1.07 to get a dollar? We divide $1.07 by 1.07 to get an answer of one.

How do we find the original price of an item before 7 per cent tax was added? Divide the price by 1.07. How do we find the price of an item before 8 per cent tax was added? Divide by 1.08. We have worked out our own method by simplifying the numbers and doing a problem backwards.

Go to extremes

Sometimes, in electronics classes, I have suggested this to students: what if the value were zero? What if it were

1 million? Often the answer is obvious in this instance. Have a go at puzzle 1; it has been around for years.

 Puzzle 1

A wildlife sanctuary has 100 beasts in the park. Some are birds that have, as you would expect, two legs each. The rest are four-legged animals. A visitor came to the park and asked a park ranger, 'How many birds and animals do you have in the park?'

The ranger said, 'We have one hundred'.

'No,' replied the visitor, 'I want to know how many animals and how many birds'.

The ranger said, 'I have never worked it out. I do know we have a total of 340 legs. I just don't know how many are two-legged birds and how many are four-legged animals'.

Could you help the visitor from this information? How can you go to extremes to solve this problem? Where can you start?

Draw a diagram or a picture

There are many times it can help to draw diagrams. Try it for the bat and balls problem we just did.

Reverse the details

Sometimes it can help to look at a problem in a different way. Can you reverse any of the details to help you solve puzzle 2?

Puzzle 2

A man is rowing a boat up a river and his hat falls in the water. He doesn't notice and keeps rowing for 10 minutes before he realises his hat is gone. He turns his boat and rows after it. He rows and retrieves his hat, which has floated 3 kilometres downstream from where it fell in the water. How fast was the river flowing?

This seems like a difficult question. We don't appear to have enough information to solve it. Is there any way we can reverse the details?

Find something you can do

Start somewhere. Is there anything you can solve? Solve it and maybe that will give you a clue for the next step.

For instance, with the bat and ball puzzle at the beginning of the chapter you couldn't answer the first two questions until you had answered question 3. Is there anything in the problem that you can solve?

Look for analogies

Can you think of a similar situation or problem that will help? Is this like a similar problem you have solved before? Is this situation similar to a situation somewhere else?

Visualise the problem

What combination would reach higher, a tall man standing on the shoulders of a short man, or would they reach higher by changing places? Picture in your mind a short man reaching on the shoulders of a tall man, and then picture a tall man reaching on the shoulders of a short man. Who is reaching higher?

Make no assumptions

After reading a problem, go back to the beginning if necessary and question what you know. Do you really know it? Are you sure what you know is correct?

Substitute

Use different terms. Take out (or add) emotional elements. For example, what if it were us, China, Iceland, your mother?

Do what high achievers do

What would you do if you could solve the problem? Look for a place to start. What would a high achiever do? Find somewhere to begin. Even if you can't convince yourself you are capable of solving a problem, at least tell yourself you will do what the high achievers would do and act as if you could.

Look for trends

Let's look for trends in two easy examples.

If you multiply a number by 10 it becomes larger, right? For example, 2 times 10 is 20. If you multiply by 5 it becomes larger, but not so large as multiplying by 10. Two times 5 is 10. If you multiply by 1 the answer is lower than if you multiply by 5. Two times 1 is 2.

If we take this trend further, it can illustrate an important principle. Let's review what we have just seen. If we multiply by a large number we get a large answer. If we multiply by a smaller number we get a smaller answer. If we drop the number we are multiplying by down to 1, the number we are multiplying stays the same.

What if we multiply by an even smaller number, like one half? A half of 2 is 1 (to say a half of a number is the same as saying we multiply the number by a half). Let's go even smaller. Multiplying by one quarter will give an even smaller answer. A quarter of 2 is one half.

We have found a trend with multiplication. Multiplying by numbers larger than one will increase the number we are multiplying. Multiplying by numbers smaller than one will decrease the number we are multiplying.

Let's see if this works for division. If we divide a number by 10, say 20, we get a smaller answer. For instance, 20 divided by 10 gives us 2. If we divide the same number by 5, the answer is not so small. So 20 divided by 5 gives us 4. If we divide the number by 1, the number remains unchanged, so 20 divided by 1 gives us 20.

If we keep lowering the number we are dividing by, what happens? To choose a number less than 1, we could choose one half. If you divide 20 apples in half, how many halves will you have? The answer is 40. Two cakes divided into quarters gives 8 pieces.

Here are two trends we have discovered that can help improve our understanding of maths.

If we are multiplying a number, the larger the number we multiply by, the larger the answer. The smaller the number we multiply by, the smaller the answer.

We know that any number multiplied by 1 stays the same, so any number multiplied by less than 1 must give a lower answer than the number we are multiplying.

There is an important warning, however. Not all trends are valid. Some maths calculations go down towards zero and then swing up again. Play and experiment with your trend to see if it is valid. Don't make up your mind too soon when you find a trend; you may be mistaken. Which brings us to the next rule.

Keep an open mind

The puzzle you are doing might look like another puzzle you have already solved, but that could be a ruse. Also, as we have just seen, not all apparent trends are valid trends. Puzzles 3 and 4 are good examples of puzzles that deliberately set out to trick you.

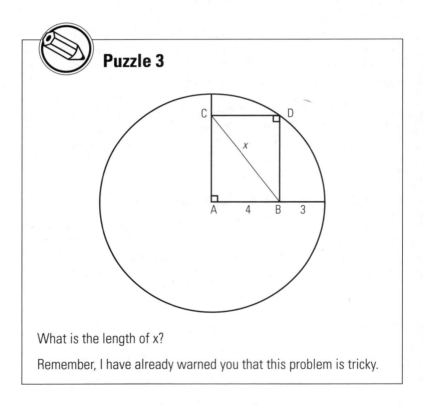

Puzzle 3

What is the length of x?

Remember, I have already warned you that this problem is tricky.

Puzzle 4

When I was young, I remember my grandmother giving me this puzzle.

A man is looking at a portrait, and he turns to his friends and says, 'Brothers and sisters have I none, but this man's father is my father's son'.

Who is the man looking at? Whose picture is it?

Understand the problem

Ask yourself, what am I being told? What am I being asked? What does this mean? Use the previous rules to try to help your understanding. Break the problem into parts to try to understand what the question is asking. Draw diagrams. Try anything to help you understand the problem.

Think outside the square

I believe the term 'thinking outside the square' originated from puzzle 5. Thinking outside the square applies to thinking outside self-imposed boundaries.

Ask yourself, 'Do I have to do it this way?'

Puzzle 5

Here is a pattern made up of nine dots. Can you make four straight lines without lifting your pencil that pass through each of the dots?

● ● ●

● ● ●

● ● ●

Puzzle 5 reminds me of puzzle 6, the old envelope puzzle we used to try to solve when I was a kid.

 Puzzle 6

Can you draw the following diagram without lifting your pencil or going back over a line?

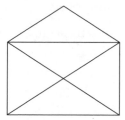

Trial and error

Trial and error may not seem like a valid strategy, but it is. Try various values to calculate. What happens? What is the trend? Do the answers give clues to the solution?

In fact, trial and error is the only way to solve some logic problems. For instance, if you are trying to solve a cryptogram where the letters in a puzzle have all been substituted and the letter W signifies a one-letter word, that word could be only 'I' or 'a'. You would try substituting the letter W with one letter and then the other to see which letter makes sense.

Try puzzle 7. The letters have been randomly changed. The only rule with cryptograms is that no letter can equal itself. See if you can solve it using logic.

 Puzzle 7

WMR RLCYRN WMR FRWMSK TSX XCR WS CSPOR L BNSEPRF,
WMR DLCWRN TSX GYPP CSPOR YW GYWM PRCC QMLVQR SD
FLJYVZ L FYCWLJR.

Here are some hints for solving cryptograms. There are only two one-letter words in the English language (I and a). The most common letters in English are E, T, A, I, S, O, N, R and H. One of the most common words in English is 'the'. 'The' includes two of the most common letters. 'Th' is also a common combination; look out for the words 'that' and 'there'.

Give the puzzle a good try before you look up the answer at the back. You really can solve it.

Practise

The way to improve your performance is to practise. Practise solving puzzles and many of these strategies will come naturally to you. You can also develop your own strategies. Buy puzzle books and try these methods to solve as many puzzles as you can. Success may be slow to begin with, but you will have a lot of fun along the way as you improve. As you solve puzzles, you will start to recognise patterns and create some of your own strategies.

Have a go at puzzles 8, 9 and 10.

 Puzzle 8

Below are four words:

WESLEY

DENMARK

ENGLAND

REMAIN

We can encode them so that all of the letters have been changed. For example, if we change T to D in one word, it is D in the other words as well. The same code must be used for all four words.

Can you work out which word is which? The order has been changed.

WESLEY	CMBOHTA
DENMARK	MBQWHBC
ENGLAND	TMOHZB
REMAIN	RMKWMS

How would you begin to solve this problem? Is there anything you can do? What things could you look for?

Puzzle 9

Jim is twice as old as his sister, Sue, was one year ago. Next year Sue will have her tenth birthday. She has already had her birthday this year. How old is Jim?

Would it help to draw a diagram? Where could you begin?

Puzzle 10

Adam is two years younger than Sarah. Their father is four times as old as Sarah will be next year. Last year, their father was the same age as their mother is now. Their mother had her 35th birthday last week.

How old is Adam?

How old Sarah?

How old is their father?

How old is their mother?

Where would you begin? Would you try to answer the first question first?

Create your own puzzles

You can make up similar problems of your own to give to your children. Children can make their own problems to puzzle their friends. This is also good practice.

Key points

- Playing with puzzles will help your child develop problem-solving skills.

- If you can't solve a problem, try a different strategy.

- If you don't know where to begin, ask yourself some questions that will help you find a starting point.

- You have a much better chance of solving a problem if you believe you will be able to solve it.

Chapter 15
Alphametics—fun with numbers

No-one can really understand mathematics without understanding the properties of numbers. The better you understand how numbers work, the better you will be able to work with them, and to solve problems involving numbers. In this chapter we take a look at alphametics—mathematical calculations that use letters rather than numbers. Alphametic problems are a fun, and challenging, way to learn the properties of numbers.

Properties of numbers

In the past, arithmetic was usually taught as a series of rules. Students learnt to solve mathematical problems by rote learning—that is, they would be taught to memorise their times tables without being given an explanation of the principles. Students who are taught using this approach usually have difficulty solving so-called word problems because they have to work out the method for themselves.

If you can teach your children to understand the properties of numbers, they will have a huge advantage over their classmates.

For instance, what are some properties of the number 5?

You could say that 5 equals 4 plus 1, 3 plus 2, or 10 divided by 2 or half of 10. You could also say that 5 is 6 minus 1 or 7 minus 2. You should know that 5 is an odd number, and any number multiplied by 5 ends in either 5 or zero.

What about the number 2? If you multiply any number by 2, you will get an even number for the answer. In other words, the units digit of the answer will be even. That is a property of 2.

How would you determine the correct answer to the following problem in a multiple-choice test?

$3473 \times 1234 =$

(a) 4 276 884

(b) 4 285 682

(c) 4 312 516

(d) 4 168 217

Do you have to calculate the answer to the problem to determine which answer is correct? No. You can simply multiply the units digits of the problem to determine the units digit of the answer. The units digit of 3473 is 3. The units digit of 1234 is 4. Three times 4 is 12. The units digit of 12 is 2, so the answer to the full problem must end in 2. The only choice ending in 2 is (b), so you have the answer without doing the whole calculation.

Of course, you could also have cast out the nines to determine the correct answer but this method is much easier.

Tip

An odd number multiplied by an odd number will give an odd number answer. An even number multiplied by an odd number will give an even answer, as will an even number multiplied by an even number. Try to work out why this is so (at the end of the chapter I provide an explanation).

We will now look at some puzzles that are a fun way to learn about the properties of numbers.

What are alphametic puzzles?

Alphametic puzzles have been around for many years and were well established in India and China more than a thousand years ago. The puzzle is a simple mathematical calculation—addition, subtraction, multiplication or division—but with the digits changed into letters of the alphabet. Each letter represents one digit; each time the letter occurs in the puzzle it represents the same digit.

The first time you encounter these puzzles they can seem strange and difficult. Often people are intimidated because they see no way to begin. With just a little practice the puzzles become both easy and challenging.

Alphametics are an entertaining way to learn the properties of numbers and to learn the basic principles of mathematics. Rote learning will not help you to solve these problems. To solve these problems, you are not only forced to use your knowledge of how numbers work, but you are also forced to think mathematically.

Most mathematical competitions in Australian schools feature at least one example of these problems to test a student's ability to think logically and his or her understanding of the properties of numbers. The test papers aren't checking how much a student has learnt or how well she worked through the year. The testers want to learn how much she understands and whether she is able to give creative answers to the questions. Playing with these puzzles is a great way to develop these skills.

The puzzles were first called cryptarithms. Alphametic is a word coined by JAH Hunter in 1955 to describe cryptarithm puzzles that consisted of actual words. Cryptarithms and alphametic puzzles are a fun way to become acquainted with the properties of numbers, as well as a way to develop thinking skills. I will introduce you gradually and easily to the puzzles and explain the thinking involved.

Let's have a go at this puzzle:

$$\begin{array}{r} A \\ B\,+ \\ \hline CD \end{array}$$

It is, in fact, impossible to solve the problem, but there is one digit we can definitely know. The highest possible values for A and B are 9 and 8. Add 9 plus 8 and you get an answer of 17. This tells us that the tens digit, C, must have a value of 1. When you are adding two numbers, the only number you can carry is 1. C is a carried number so it can only be 1.

Let's take a look at the following puzzle.

```
 AB      or      AB
 ×2              AB +
BCC             BCC
```

They are both the same calculation.

I remember my mathematics teacher, Harry Forecast, saying that solving mathematical problems is like playing Sherlock Holmes. You have to look at the clues and draw your conclusions. What conclusions can we reach from this puzzle?

There is a carried number, B. This tells us that B must be 1. We can add B + B (1 + 1) in the units column to find that C must equal 2. The problem now looks like this:

```
 A1
 A1 +
122
```

We have the answer to B and C; we need only to find A. A plus A equals 12 (BC) so A must be 6.

This gives us all of the values so we can write the answer in full:

```
 61
 61 +
122
```

Try the following puzzles. The solutions are provided in appendix B.

Puzzle 1

Here is a very easy puzzle to begin with. The frog says, 'I swallowed two bees and became ill'. Here is how we write the puzzle:

```
    I
  BB +
   III
  ILL
```

Puzzle 2

```
  ABC
  × 2
  CDDB
```

Puzzle 3

```
  ABA
  ABC +
  CCDB
```

Puzzle 4

```
  AS
  A +
  MUM
```

Puzzle 5

```
   I
  DID +
  TOO
```

Puzzle 6

```
  ABCD
  EBCD +
  EFEAA
```

Puzzle 7

```
  LOSE
  SEAL +
  SALES
```

Puzzle 8

Here is the first known example of a cryptarithm and the first that I encountered. This puzzle introduces some more principles to consider. If you get stuck with the problem, don't look up the answer but simply put it aside for a while and come back to it later. Make notes about the value of each letter. Solving these problems is a skill that can be developed.

$$SEND$$
$$\underline{MORE\,+}$$
$$MONEY$$

Puzzle 9

My brother-in-law presented this problem to me when I was quite young. It is one of my favourites.

$$ABCDE$$
$$\underline{\times\,4}$$
$$EDCBA$$

Remember that all numbers multiplied by 2 (or an even number) must give an answer of an even number. (The final digit must be even.)

Trying to solve these puzzles will force you to think about how numbers work and the properties of numbers. It is a fun way to learn and very satisfying if you get the correct answer.

Explaining odds and evens answers

Now, to answer the question posed at the beginning of the chapter, why is it that if you multiply an odd number by an odd number the answer is an odd number, and an odd number multiplied by an even number or an even number multiplied by an even number produce an even number? An even number is a multiple of 2—that is, it can be evenly divided by 2, or one of its factors must be 2. So multiplying any number by an even number is the same as multiplying it by 2 so far as the odd or even outcome is concerned. For instance, if you multiply any number by 6 you are multiplying it by 3 times 2.

That takes care of multiplying an even number by an even number and an odd number by an even number.

What about multiplying an odd number by an odd number? Any odd number is one more than an even number. For instance, 5 is an odd number. It is one more than 4, an even number. So, if you multiply an odd number by an odd number you are multiplying it by an even number and then adding the odd number to the answer.

Let's see how this works with 7 times 5. Five is equal to 4 plus 1. Seven times 4 is 28, plus one more 7 makes five sevens. Adding an odd number to an even number will give an odd answer.

Key points

- Alphametics are a fun way to learn the properties of numbers.
- Solving alphametic problems develops other thinking skills.

Chapter 16
Encyclopaedia salesman puzzle

This is one of my favourite puzzles. It illustrates a whole lot of the strategies for solving problems when you have no idea where to start.

I have given this puzzle to students all around the world. The calculations are not difficult; a grade four or five primary school student should be able to make the calculations, but very few people are able to solve the puzzle.

I was teaching a class of engineering students and they knew I couldn't resist solving puzzles. Often a class would leave a puzzle on the board and I wouldn't begin my class until we had solved it or, at least taken it down for homework. A student came to me with this puzzle written on a sheet of paper.

An encyclopaedia salesman came to a house and knocked on the door. He explained to the lady why she needed a set of his encyclopaedias.

The lady said, 'I will buy them if you can tell me the ages of my three children. If you multiply the three ages together you get an answer of 36'.

The salesman thought for a moment and then said, 'You will have to give me another clue'.

The lady said, 'All right, if you add the ages together, they add up to our house number'.

The salesman had a look at the number of her house, scratched his head, and then said, 'You will have to give me another clue'.

The lady said, 'Okay, but this is the last help you will get. The youngest child has red hair'.

That was all the salesman needed. He told the lady the ages of her children and walked away counting his money from the sale.

Three questions to answer

Here are three questions for you to answer:

1   What were the ages of the three children?

2   What was the number of the lady's house?

3   What difference did the colour of the child's hair make? Without the final clue the salesman could only guess the ages. Why did he need this final clue?

My heart sank when I read the puzzle. I am not sure if the student saw the expression on my face but he said, 'Imagine you are the salesman'.

I gave the class an assignment to keep them busy and went to work on the puzzle. The student's statement to imagine I was the salesman gave me the start I needed.

This is a classic example of a puzzle that intimidates. You are not given the lady's house number, and you wonder how did the last clue help? How did knowing the colour of the youngest child's hair help in the least?

Most people never even begin to solve the puzzle or they make a wild guess. I have had telephone calls in the middle of the night from families asking if they have the right answer. The simple answer to that question is this: if you have to ask then you haven't solved it.

Let's look at some aspects of the problem and how we should set about solving it.

First, find something you *can* do: find somewhere to start. I began by asking myself, what would I do if I were the salesman?

Taking the first step may give you the clue to the next step.

Try to solve the problem by yourself. If you can't, look at the hints below on where you could begin before you turn to appendix C for the answers and explanation.

Hints for the puzzle

This is a puzzle that intimidates. People have got angry when I have given it to them. One man said, 'You are just making fun of us'.

Obviously, it is the red hair that makes you think it is not a genuine puzzle. How can the colour of the child's hair help solve the children's ages?

This is a classic puzzle to illustrate what we must do if we have no idea how to solve a problem. Use the strategies taught in this book and you will discover you can solve this apparently impossible problem.

The first suggestion I will give is to look for something you *can* do. The student who gave me the puzzle helped when he said, 'Imagine you are the salesman'.

I thought that the salesman, knowing he had a big commission riding on making the sale, would start looking for some answers.

I followed my rules on problem solving and was embarrassed afterwards to think I had missed some obvious lines of thought. There were some questions I should have asked myself.

1   Why couldn't the salesman answer after the first clue?

2   Why couldn't he answer after the second clue?

I did ask myself the question, where would the salesman begin?

Is there anything you can do that the salesman would be doing?

Now, spend some more time on the problem before you give up. I assure you that you are capable of solving this puzzle. (You will find the answer in appendix C.)

Key points

- What can you do to start?

- What would the salesman do? Where would he begin?

- Imagine you are the salesman and you are desperate to make the sale.

Part III

Solutions and practice sheets

In this section I not only give the answers to the puzzles, but I also explain the thinking involved in solving them.

 # Appendix A

Solutions to puzzles in chapter 14

Puzzle 1

The standard way to solve this problem is to use simultaneous equations, but there is an easier way. Let's say that all the beasts are birds or they are all animals. If the 100 beasts were birds, how many legs would they have? One hundred times two legs would make 200 legs. But we have 340 legs in the problem. That is 140 too many. They must belong to the animals. Giving two more legs to each of the four-legged animals, we have 140 legs to give them. If each four-legged animal receives another two legs, how many animals can we supply? Half of 140 is 70, so 70 animals receive an extra pair of legs to make up the 140. That means there are 70 four-legged animals. How many beasts are there all together? One hundred.

If 70 are four-legged animals, that leaves 30 that must be two-legged birds. How does this answer compare with the information we were given?

$$70 \times 4 = 280$$
$$30 \times 2 = 60$$
Total **= 340**

We were told we had 340 legs, so the answer checks with our information.

When would you use this in practice?

Say you are in a band and you perform at an after-school concert. Admission is $10 per person, but students from your school only have to pay $5. Two hundred people attend and your takings are $1400.

How many students from your school attended? We will use the same method.

Let's assume all who attended were from your school. Two hundred times $5 is $1000. You received $400 more than that. That $400 was paid by people who weren't students from your school. How much more did the non-students pay per person? Five dollars. So divide $400 by 5 to find out how many paid $5 extra (non-students). Four hundred dollars divided by 5 is 80, therefore there were 80 non-students. There were a total of 200 people. How many were students from your school? Two hundred minus the 80 non-students leaves 120 from your school.

Does this check with our information?

$$80 \times 10 = 800$$
$$120 \times 5 = 600$$
Total **= 1400**

176

Yes, the answer checks with our information, because your takings were $1400.

This is an example of going to extremes. We assumed a maximum number of students attended, and then used this assumption to solve the problem.

Puzzle 2

Let's draw a diagram and see if that will help.

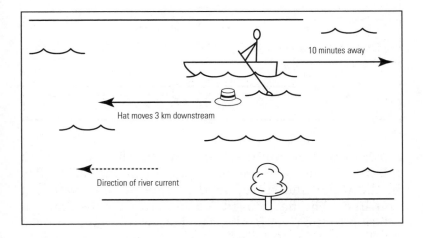

10 minutes away

Hat moves 3 km downstream

Direction of river current

It doesn't seem to help much. Can we make the problem simpler?

What if we have the water in the river standing still and the river banks moving instead? How would that affect the problem?

We would see the hat fall into the water and stay still. The man rows the boat for ten minutes, then finds his hat is missing. He turns the boat and rows back to where he left his hat. How long did the man row away from the hat for? Ten minutes. How long does it take him to row back to where it fell? (The water is stationary, remember.) Ten minutes. How long did it take him altogether to row in each direction? Two times 10 minutes, or 20 minutes.

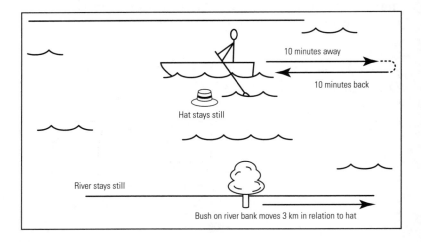

10 minutes away

10 minutes back

Hat stays still

River stays still

Bush on river bank moves 3 km in relation to hat

Now, in that 20 minutes, how far did the hat move in relation to the river banks? (Remember, the hat landed in the water and stayed in the same piece of water, if you like.) The hat moved 3 kilometres downstream in relation to the river banks. The hat, or the water, moved 3 kilometres in 20 minutes. How far would it move in an hour? Nine kilometres (20 minutes divides evenly 3 times into an hour, 60 minutes. So in an hour, it would cover 3 times the distance, or 3 times 3 kilometres). The current is flowing at 9 kilometres per hour.

Puzzle 3

This is my favourite deceptive puzzle. I have annoyed many mathematicians with this problem.

Most people try to solve the puzzle using Pythagorus's theorem. They miss the fact that the length of the line B to C is the same as from A to D. Take another look and you will see that A to D is the radius of the circle. Therefore, A to D is the same as B to C, also the radius of the circle. A to D is equal to 4 plus 3. If you draw one more line on the diagram, all becomes clear.

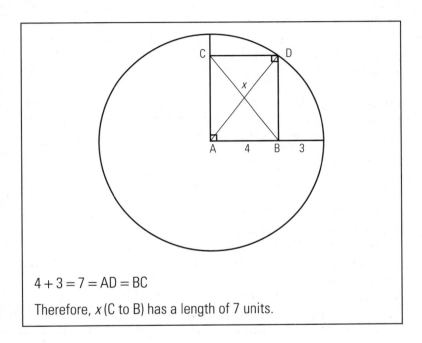

$4 + 3 = 7 = AD = BC$

Therefore, x (C to B) has a length of 7 units.

Puzzle 4

Before you read any further, try to solve the problem for yourself. Take the problem apart. Draw a diagram. Solve the problem one part at a time. Do you have the answer? If you said, 'He is looking at a picture of himself', you have given the answer most people give. And they are wrong! He is looking at a picture of his son.

Marilyn vos Savant tells the story in her book, *Brain Building*, of how she published this problem in her column in *Parade* magazine. Vos Savant had the highest IQ ever recorded according to *The Guinness Book of World Records*, so she is not someone you would quickly disagree with. When she published the answer, she received a flood of angry letters saying she was wrong: the man was looking at a portrait of himself. She printed some letters as she received them, telling her they had been told by their professor or they had the problem in an entrance exam and they had been told he was looking at himself. They demanded a retraction, an apology and a correction in her column. What intrigued me was that no-one said, 'I have worked it out and this is why you are wrong'. They quoted authorities. Remember, the answer you have been given may not necessarily be correct.

A diagram can help with this puzzle.

We see the speaker looking at the picture.

He says, 'This man's father ... '. So we draw a line up from the portrait to the father.

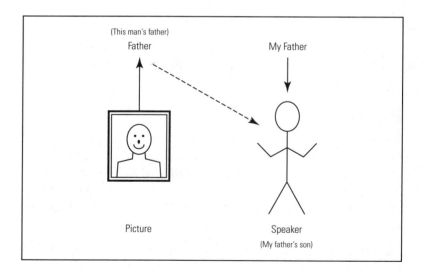

He goes on to say, '…is my father's…'. Draw a line from the speaker to denote his father, '…son'. Who is his father's son? He is, because he had no brothers or sisters. Write on the diagram that he is his father's son. Now, who is this man's father? The speaker, who is 'my father's son'. Therefore, the speaker is the father of the man in the picture. You could draw a line connecting the father with my father's son (the speaker) because 'this man's father is my father's son'.

Here is another way to look at the problem. Ask yourself questions that may shed light on the answer:

- Who is this man's father?

  My father's son.

- Who is my father's son?

  I am (I have no brothers or sisters).

- If I am this man's father, who is this man?

  My son.

By the way, I was also taught that the correct answer is that he is looking at his own picture. I wonder if the wording has been changed or if there are two versions of the puzzle.

Puzzle 5

If you took notice of the heading, 'Think outside the square', you may have had no difficulty solving this puzzle because you have to literally draw the lines outside and beyond the square. Then the answer is easy (shown overleaf).

All too often we place our own restrictions on problems we are trying to solve, whether they are mathematical, career or personal. Or we place other people's restrictions on ourselves. Check first to see if the restriction is valid.

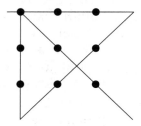

Puzzle 6

I solved this puzzle by looking for two lines that must join. For instance, at the bottom left corner of the diagram there are three lines that connect. The only connection that is certain is the roof. I began with the roof and then decided that there were two options from there — a line could be drawn either down or across. I tried down first and found that it led nowhere, so I drew a line across. It went easily from there but I was unable to draw the final wall. I saw that I could fix that by drawing the wall first and continuing on to the roof.

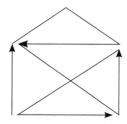

You may find the answer using hit and miss methods but it is more satisfying to work logically. This puzzle kept the kids in my neighbourhood busy for weeks while we tried to solve it.

Puzzle 7

WMR RLCYRN WMR FRWMSK TSX XCR WS CSPOR L
BNSEPRF, WMR DLCWRN TSX GYPP CSPOR YW GYWM PRCC
QMLVQR SD FLJYVZ L FYCWLJR.

The first word of the cryptogram is 'the'. The letters W
and R occur frequently. The combination WM is also
common. The one-letter word L must be either 'a' or 'I'.
The solution to the cryptogram can be found throughout
this book — 'The easier the method you use to solve a
problem, the easier you will solve it with less chance of
making a mistake'.

Solving cryptograms will improve your spelling and
your thinking skills. A number of books and magazines
include cryptograms for you to play with. I did a Google
search for cryptograms and found a number of sites that
offer them for free.

Puzzle 8

WESLEY	CMBOHTA
DENMARK	MBQWHBC
ENGLAND	TMOHZB
REMAIN	RMKWMS

How would you begin to solve the problem? Is there
anything you could do? How could you make a start?
What things could you look for?

The first strategy might be to count the number of letters
in each word — WESLEY has six letters, DENMARK has
seven letters, ENGLAND has seven letters and REMAIN
has six letters.

CMBOHTA has seven letters, MBQWHBC has seven letters, TMOHZB has six letters and RMKWMS has six letters.

Next we could look for double letters or recurring letters. Wesley has an E for the second letter and the second last letter. England has N as the second letter and the second last letter. Denmark begins with D and England ends with D. That should be enough information to give us the final answer.

Puzzle 9

Where can we begin?

We read that Sue will be 10 next year. How old is she now? Nine years old. Jim is twice as old as his sister Sue was one year ago. How old was Sue one year ago? She is 9 now, a year ago she was 8. How old is Jim? Twice 8 is 16. Jim is 16 years old.

Puzzle 10

How old is Adam?

How old is Sarah?

How old is their father?

How old is their mother?

Can we answer any of the questions from the information we have? Yes, we read the mother had her 35th birthday last week. We can answer the last question. Their mother is 35.

Last year, their father was the same age as their mother is now. This tells us that the father was 35 last year.

How old is he now? Thirty-six. That's the third question answered.

Their father is 4 times as old as Sarah will be next year. How old is the father? Thirty-six. How old will Sarah be next year? A quarter of 36, or 9 years old. If she will be 9 next year, how old is she now? Eight years old. That's the second question answered.

Adam is 2 years younger than Sarah. Adam is 6 years old. You'll notice we had to answer the questions in reverse order.

We couldn't answer any questions until the last question was answered. Sometimes we have to answer a lot of questions before we can get to the question we want to answer.

Appendix B

Solutions to alphametic problems in chapter 15

Alphametics are a fun way to learn the properties of numbers. Following are the solutions to the puzzles in chapter 15.

Puzzle 1

```
   I
  BB +
  ILL
```

Where do we begin? Here is a clue. Ask yourself, what is the hundreds digit? If you add a single-digit number to a two-digit number, how many hundreds can you have in the answer? This is an important principle for solving all addition alphametic problems.

If you take any two-digit number, say 77, and add to it a single-digit number, how many hundreds can you expect in the answer? If BB did in fact represent 77 it would be impossible to add a single-digit number and get a three-digit answer. That statement should give you the clue to solve the problem.

Also consider, the first digit in the answer (the hundreds digit) is a carried number. When you are adding two numbers there is only one number you can carry.

So, what two-digit number when added to a single-digit number gives an answer in the hundreds? The only possible answer for the hundreds digit is one. You cannot add a two-digit number to a one-digit number and get an answer of three or four hundred.

Now, what two-digit number plus one gives an answer of at least one hundred? Ninety-nine.

The answer is clear:

$$\begin{array}{r} 1 \\ 99\,+ \\ \hline 100 \end{array}$$

Puzzle 2

$$\begin{array}{r} ABC \\ \times\,2 \\ \hline CDDB \end{array}$$

This is the same as:

$$\begin{array}{r} ABC \\ ABC\,+ \\ \hline CDDB \end{array}$$

We can begin with the first letter of the answer, C. It is a carried digit. The only digit you can carry when adding two digits is 1. The highest possible value for A is 9 (A plus A is 9 plus 9 equals 18). You would write the 8 and carry the 1. We don't know the value of A, but A plus A gives a two-digit answer. The first digit of the answer can only be 1, so C equals 1.

We can substitute 1 for C in the units column. Now that we know what C equals, 1 plus 1 equals 2, so B must be 2.

This means we can work out what D is. We know that B is 2 so D must be 4. The problem so far looks like this:

```
 A21
 A21 +
1442
```

That only leaves us with A to find. A plus A equals 14 so A must be 7.

The problem is solved:

```
 721
  ×2
1442
```

Puzzle 3

```
 ABA
 ABC +
CCDB
```

The first digit is a carried number. The only number you can carry when adding two numbers is 1, so C equals 1.

Therefore, A plus A gives an answer of 11. However, this is impossible, so we must have 1 carried from the previous addition of B plus B.

If A plus A plus 1 carried gives us 11, then A must be 5. (If A plus A plus 1 equals 11, then A plus A must equal 10. If A plus A is 10, then A must be 5.)

We now move to the units column. We know the values of A and C. A is 5 and C is 1. The problem so far looks like this:

```
  5B5
  5B1 +
11DB
```

Five plus 1 is 6, so B equals 6.

In the tens column, B plus B equals D with one carried. We know B is 6, so 6 plus 6 equals 12. D must be 2 with the 1 carried to the hundreds column.

The answer is:

```
  565
+ 561
 1126
```

Puzzle 4

```
  AS
   A +
MUM
```

We are confronted with something a little different here.

The first digit of the answer is a carried digit, so it must be 1. A plus 1 carried gives a two-digit answer. A must be 9; no other digit plus 1 (carried) will give a two-digit answer. So, if A is 9, plus 1 carried makes U, then U must be zero (9 + 1 = 10). The problem so far looks like this:

```
 9S
  9 +
101
```

Now to the units column. S plus A must equal 11 (M is 1 and we carried 1 to the tens column). Two plus 9 equals 11, so S must equal 2.

```
 92
  9 +
101
```

Puzzle 5

```
  I
DID +
TOO
```

The first step with this problem is to work on the tens and units digits because the T is not a carried number. We see from the tens digits that I plus 1 carried equals O plus 10. In the units column I plus D equals O plus 10.

Again, looking at the tens digits, I plus 1 carried equals a two-digit number. The only digit plus 1 to give a two-digit answer is 9. Nine plus 1 equals 10. I must equal 9 and O must equal zero. The problem so far looks like this:

```
  9
D9D +
TOO
```

We return to the units digits. If I is 9 and O is zero, D can only be 1.

In the tens column, we have I plus 1 (carried) equals 10. In the units column we have I plus D equals 10. Because we know that I is 9 then D has to be 1.

In the hundreds column we have D plus 1 carried equals T. D is 1 so T must be 2.

The problem is solved:

$$
\begin{array}{r}
9 \\
191\, + \\
\hline
200
\end{array}
$$

Puzzle 6

$$
\begin{array}{r}
ABCD \\
EBCD\, + \\
\hline
EFEAA
\end{array}
$$

We begin with the first digit of the answer, E, which is carried, so it must be a 1.

A plus 1 equals a two-digit number. The highest A can be is 9, so 9 plus 1 is 10 or, if 1 is carried from the previous column, we get an answer of 11.

F can't be 1 because E already equals 1, so F has to be zero.

In the units column we see that D plus D equals A. A must be even. We can't add a number to itself and get an odd number for the answer. Another way of seeing it would be to say that D multiplied by 2 must give an even answer. So we know A is even.

We already saw that A plus 1 is 10. Because A is even, it can't be higher than 8. So there must be another 1 carried from the previous column. That will work because 8 plus 1 plus 1 carried equals 10. A equals 8. The problem so far looks like this:

```
  8BCD
  1BCD +
101¹88
```

B plus B must equal 10 because 1 carried from the previous column gives us 11. (We already determined that 1 was carried to add to 8 plus 1.) If B plus B equals 10, then B must be 5.

We can also see that D plus D in the units column gives an answer of 8. Could it be 18? No, because that would mean carrying 1 to the tens column and having an odd digit for the answer. So, D must be 4:

```
   85C4
   15C4 +
10¹1¹88
```

All that is left to solve is C. C plus C equals 8 with 1 carried, meaning C plus C is 18. C must be 9. The problem is solved:

```
   8594
   1594 +
10 ¹1¹88
```

Puzzle 7

```
 LOSE
 SEAL +
SALES
```

We begin with the first digit of the answer, S. S is a carried number so it can only be 1. We now have:

```
 LO1E
 1EOL +
1ALE1
```

We see that L plus 1 gives us 1A. The highest possible value for L is 9. The only possibilities are 9 plus 1 equals 10 or with 1 carried we would get 11. We already have the value for 1 so A must be zero.

If L were 8 we would have 8 plus 1 equals 9, plus 1 carried equals 10. Either way we get 10 for an answer so A is definitely zero.

In the tens column, we have 1 plus zero equals E. We must have a carried 1 to add, otherwise 1 plus zero equals 1 instead of E.

If 1 plus 1 carried equals E, then E must be 2. The problem so far looks like this:

```
 LO12
 120L +
10L2¹1
```

That leaves O and L to be solved. In the units column, 2 plus L equals 11, so L must be 9.

```
  9012
  1209 +
10092¹1
```

O plus 2 equals 9, so O must be 7. The problem is solved:

```
  9712
  1209 +
10 92¹1
```

If you had trouble solving that puzzle, read through the explanation again. The reasoning is not difficult for each step. The real problem is not understanding the reasoning but finding each step for yourself. That will come with practice.

Puzzle 8

```
 SEND
 MORE +
MONEY
```

There are similarities here with the previous puzzle. A lot of the reasoning is identical. If you haven't solved it, go back and have another try.

M is a carried number so it has a value of 1.

S plus 1 is 10. If we had carried a number to add (the only number we can carry is 1), then we would have 11. We already know that M is 1 so O must be something else. The only other possible value is zero.

We now have:

```
  SEND
 1ORE +
 1ONEY
```

We have S as either 8 or 9, the only possible values to give an answer of 10.

We can see that E plus zero makes N, so there must be a carried 1 from the previous column; therefore, E plus 1 equals N.

Now we know that N is one more than E, we can see in the tens column that adding R to N gives us an answer that is 1 less.

What number do you add to any digit to get a units digit that is 1 less. That is like asking, what number do we add to 5 to get an answer of 14? The answer is 9. R must either equal 9, or 8 with 1 carried.

We can reason further with E plus zero equals N. We know that S and R are the only possible values for 8 and 9, so there is no carried number to add to S plus 1 equals 10. That means that S is 9 and R is 8. The problem now looks like this:

```
   9END
  1O8E +
 1ON¹E¹Y
```

Now we have to do some different thinking. The highest possible values for D and E in the units column are 6 and 7, because we already have values for 8 and 9. However, 6 and 7 are not possible because N is one more than E and all of the higher values are taken. So the highest values for D and E are 5 and 7. In fact, they are the only possible values because together they equal 12, which means Y is 2. Anything lower for Y is impossible, because we already have zero and 1 allocated.

Therefore, E, N and D must equal 5, 6 and 7, not necessarily in that order. E can't be 7 because N is E plus 1. E must be 5 or 6. We have seen that D and E are 5 and 7. E isn't 7 so it must be 5. N is E plus 1 so N is 6. D must be 7. The finished problem looks like this:

$$
\begin{array}{r}
9567 \\
1085 + \\
\hline
106^15^12
\end{array}
$$

That was a tough one if you have never done these before. Most of my students find these problems difficult to begin, then, with practice, they start to like them and ask for more to solve.

Puzzle 9

$$
\begin{array}{r}
ABCDE \\
\times 4 \\
\hline
EDCBA
\end{array}
$$

Although this puzzle may seem difficult, we do have some clues. The final digit of the answer is an even number. How do we know? Because any number multiplied by 2 must have an even number for an answer, and 4 is 2 times 2. So, we know that A is even.

What else can we tell about A? A multiplied by 4 gives a one-digit answer. What even number multiplied by 4 gives a single-digit answer? The only possibility is 2. A must be 2.

What is E? A multiplied by 4 is 8. One more carried would give us 9. But, in the units column, E times 4 gives an answer ending in 2. Eight times 4 is 32 but 9 times 4 is 36. E must be 8.

Let's fill them in.

$$\begin{array}{r} 2BCD8 \\ \times\,4 \\ \hline 8DCB^{3}2 \end{array}$$

What else can we see? B times 4 gives a single-digit answer because nothing has been carried to the final column. B must be 1. It can't be zero because of its position in the answer. D times 4 must give an even answer, plus the 3 carried must give an odd digit for B. So B equals 1. The problem looks like this, so far:

$$\begin{array}{r} 21CD8 \\ \times\,4 \\ \hline 8DC1^{3}2 \end{array}$$

D times 4, plus 3, gives an answer ending in 1. Don't add the 3 and D times 4 gives an answer ending in 8. D can't be 2 because A equals 2. D must be 7 because 7 times 4 equals 28. The problem now looks like this:

$$\begin{array}{r} 21C78 \\ \times\,4 \\ \hline 87C^{3}1^{3}2 \end{array}$$

That only leaves C. C multiplied by 4, plus 3 carried, gives an answer with the units digit equalling itself.

We can also work out how many are carried to the next column. One times 4 is 4, plus the number carried equals 7. We must have carried 3.

Now we know that C times 4 gives a number in the thirties, plus 3 gives us an answer with a units digit of C.

There are only two numbers you can multiply by 4 to give an answer in the thirties, eight and nine. We already have 8 so C must be 9.

Does it work? Nine times 4 is 36, plus 3 carried makes 39—the Cs match. Here is the final answer:

$$\begin{array}{r} 21\,978 \\ \times\,4 \\ \hline 87^39^31^32 \end{array}$$

You can find plenty of these puzzles on the internet. Do a search for 'cryptarithm' or 'alphametic'. There are also books that discuss these kinds of puzzles.

Appendix C

Solution to encyclopaedia salesman puzzle

My first thought when the student gave me the puzzle was, this is one puzzle I won't be able to solve. Then the student kindly advised me, imagine you are the salesman.

I thought to myself, if I were the salesman, and had a big commission riding on whether I was able to solve the puzzle, what would I do? The obvious place to start is to find what combinations of ages would multiply to thirty-six? Most students don't take this first step.

So, I went about finding the factors of thirty-six.

$1 \times 1 \times 36$

$1 \times 2 \times 18$

$1 \times 3 \times 12$

$1 \times 4 \times 9$

$1 \times 6 \times 6$

$2 \times 2 \times 9$

$2 \times 3 \times 6$

$3 \times 3 \times 4$

These are the only combinations of three ages that give an answer of 36 when multiplied together. There are 8 possibilities. If the salesman were to guess from these combinations he would have 1 chance in 8 of getting the answer right.

So he asked for a second clue. He was told that the ages add up to the number of the lady's house.

So what would the salesman do? He would check the house number and then add the ages. Maybe we have no idea how this would help because we aren't given the house number. So we add the ages and hope that this might give us a clue.

$$1 + 1 + 36 = 38$$

$$1 + 2 + 18 = 21$$

$$1 + 3 + 12 = 16$$

$$1 + 4 + 9 = 14$$

$$1 + 6 + 6 = 13$$

$$2 + 2 + 9 = 13$$

$$2 + 3 + 6 = 11$$

$$3 + 3 + 4 = 10$$

What is the house number? The answer stands out. Had the house number been 10 he would have given the answer without the need for another clue—there is only one combination that adds to ten. So the house number had to be 13 because that is the only answer that is duplicated.

So now the possibilities for the children's ages are:

$$1 + 6 + 6 = 13$$

$$2 + 2 + 9 = 13$$

He doesn't know which combination of ages that add to the number of the lady's house to choose, so he asks for another clue.

The lady said, 'The youngest child has red hair.'

We see that with the first combination the older children are twins and with the second the younger children are twins.

The lady is actually saying two things with her statement.

- that the child has red hair, and

- that there is a youngest child: he or she is not a twin.

Most of us, especially in the west, only see one fact from the lady's statement—the child has red hair. We miss the fact that there is a youngest child.

So the children's ages are 1, 6 and 6.

What is the house number? It has to be thirteen or the salesman wouldn't have needed another clue.

And what difference did the colour of the hair make? No difference whatsoever. The importance of the third clue was that there was a youngest child. The youngest child was not a twin.

I felt embarrassed after giving the answer that I hadn't approached the problem more logically.

I should have asked myself, why couldn't the salesman have given the answer immediately he was given the first clue? Obviously, because there are several possible answers.

Why couldn't he have given the solution after the second clue? Because more than one combination of ages would have added up to the house number. This would have explained the need for a third clue.

After I have given the answers and the explanation I have had many students tell me, 'Yes, but even with twins there is an older and a younger twin'. My answer is that we can use inductive reasoning. The lady was satisfied with the salesman's answer and bought the encyclopaedias and the salesman went away counting his money. Therefore his reasoning was right on the button.

If you solved the problem by yourself, congratulations. Don't feel bad if you couldn't get the answer; not many students do. But the puzzle is a good example of following the steps to successful problem-solving. And it is a very good example of a puzzle that intimidates you before you even begin to solve it.

Appendix D

Practice sheets

Please photocopy these sheets as required. You can also download the practice sheets in PDF form from my website so you can print them off from your computer. The website address is www.speedmathematics.com.

Go to the page for resource materials for teachers to access the downloads.

Using the practice sheets

Your child should begin with practice sheet one. You will notice that it has the answers to 3 times 3, 3 times 4 and 4 times 4. After using this practice sheet for a week the answers will no longer be needed—your child will have learnt them. This approach is easier than getting your child to calculate the answers using 5 as a reference number.

The first column consists of simple calculations that were introduced in chapter 2. At this stage, the child doesn't need to understand the concept of using a reference number (described in chapter 3). To solve the calculations in the second column the child will need to use a reference number. If you are teaching a very young child it might be a good idea to just work with the first column for a while, until the child can subtract and multiply the numbers in the circles with confidence.

I encourage children to solve practice sheets one and three around three times a week. They will quickly memorise the answers for their basic tables through repetition. They should graduate to practice sheets two and three as soon as they are comfortable with basic calculations.

Practice sheet two is similar to practice sheet one. I recommend that the child memorises, rather than calculates, the answers to the problems in the third column. I would make an exception for the problems using 5 as a multiplier. Let the child multiply the number by 10 and then halve the answer according to the method taught in chapter 6.

Sheet one

$9 \times 9 =$
◯ ◯

$9 \times 8 =$
◯ ◯

$9 \times 7 =$
◯ ◯

$9 \times 6 =$
◯ ◯

$5 \times 9 =$
◯ ◯

$4 \times 9 =$
◯ ◯

$3 \times 9 =$
◯ ◯

$8 \times 8 =$
◯ ◯

$7 \times 8 =$
◯ ◯

⑩ $6 \times 7 =$
◯ ◯

◯ $6 \times 6 =$
◯ ◯

◯ $5 \times 6 =$
◯ ◯

◯ $4 \times 8 =$
◯ ◯

◯ $4 \times 7 =$
◯ ◯

◯ $3 \times 8 =$
◯ ◯

◯ $5 \times 8 =$
◯ ◯

◯ $5 \times 7 =$
◯ ◯

◯ $5 \times 6 =$
◯ ◯

⑩ $7 \times 8 = 56$
③ ②

$3 \times 3 = 9$

$3 \times 4 = 12$

$4 \times 4 = 16$

Sheet two

$9 \times 9 =$ ○ ○	⑩ $6 \times 7 =$ ○ ○	$5 \times 5 =$
$9 \times 8 =$ ○ ○	$6 \times 6 =$ ○ ○	$4 \times 6 =$
$9 \times 7 =$ ○ ○	$5 \times 6 =$ ○ ○	$3 \times 7 =$
$9 \times 6 =$ ○ ○	$4 \times 8 =$ ○ ○	$4 \times 5 =$
$5 \times 9 =$ ○ ○	$4 \times 7 =$ ○ ○	$3 \times 6 =$
$4 \times 9 =$ ○ ○	$3 \times 8 =$ ○ ○	$3 \times 5 =$
$3 \times 9 =$ ○ ○	$5 \times 8 =$ ○ ○	
$8 \times 8 =$ ○ ○	$5 \times 7 =$ ○ ○	
$7 \times 8 =$ ○ ○	$5 \times 6 =$ ○ ○	

Sheet three

(100) 98 × 95 =
○   ○

○ 97 × 95 =
○   ○

○ 98 × 94 =
○   ○

○ 97 × 96 =
○   ○

○ 95 × 96 =
○   ○

○ 94 × 97 =
○   ○

○ 93 × 98 =
○   ○

○ 96 × 96 =
○   ○

○ 94 × 96 =
○   ○

○ 99 × 99 =
○   ○

○ 99 × 98 =
○   ○

○ 98 × 96 =
○   ○

○ 98 × 98 =
○   ○

○ 97 × 97 =
○   ○

○ 97 × 98 =
○   ○

○ 99 × 96 =
○   ○

○ 98 × 92 =
○   ○

○ 97 × 89 =
○   ○

○ 98 × 75 =
○   ○

○ 97 × 75 =
○   ○

○ 98 × 70 =
○   ○

○ 98 × 82 =
○   ○

○ 90 × 81 =
○   ○

○ 90 × 81 =
○   ○

○ 80 × 85 =
○   ○

○ 92 × 95 =
○   ○

○ 93 × 95 =
○   ○

Sheet four

○ 11 × 11 =

○ 11 × 12 =

○ 11 × 13 =

○ 11 × 14 =

○ 11 × 15 =

○ 12 × 12 =

○ 12 × 13 =

○ 12 × 14 =

○ 13 × 13 =

○ 13 × 14 =

○ 13 × 15 =

○ 13 × 16 =

○ 13 × 17 =

○ 14 × 14 =

○ 14 × 15 =

○ 14 × 16 =

○ 15 × 15 =

○ 15 × 16 =

○ 16 × 16 =

○ 16 × 17 =

○ 17 × 17 =

○ 18 × 18 =

○ 18 × 19 =

○ 19 × 19 =

Sheet five

○ 9 × 11 =

○ 9 × 12 =

○ 9 × 15 =

○ 9 × 13 =

○ 16 × 9 =

○ 14 × 9 =

○ 8 × 12 =

○ 14 × 9 =

○ 8 × 12 =

○ 12 × 7 =

○ 12 × 6 =

○ 5 × 12 =

○ 4 × 12 =

○ 8 × 13 =

○ 8 × 14 =

○ 7 × 13 =

○ 7 × 14 =

○ 6 × 13 =

○ 6 × 14 =

○ 8 × 15 =

○ 7 × 15 =

# Appendix E

Why the method works

Experience has shown that young children who learn this method excel at mathematics and understand what they are doing better than most students taught the traditional way.

Over the years, I have been asked whether children should understand my methods before they are taught them. My response is how much do children understand when they recite, six sevens are 42, seven sevens are 49, eight sevens are 56?

I was teaching a grade five class at a local primary school and was about 30 minutes into the lesson when a girl said, 'I've worked out how to calculate bigger numbers. Is this right?' She showed me how she had calculated 109 times 109 using the circles and, of course, she had the correct answer. Playing with the method had stimulated her confidence and her motivation and she made use of what she had learned in just a short time. She was extremely pleased and highly motivated. And she was proud of what she had worked out for herself.

A simple explanation

If you multiply 96 times 99 you can multiply 96 by 100 and then subtract 96 for the answer. What is the easy way to subtract 96? Subtract 100 and add 4. So, 96 times 100 is 9600. Subtract 86 by taking 100 (9500) and add the 4 and you get 9504. This is the same calculation as using the circles.

Let's take the example of 8 × 9. You could multiply by 10 and then subtract 8 because you have multiplied it by one too many.

Ten times 8 is 80. Subtract 8 and you get 72, the answer. What is the easy way to subtract 8? Subtract 10 and add 2.

Now let's look at 7 × 8. You can multiply 7 by 10 and subtract 2 times 7.

So, 7 times 10 is 70. Subtract 2 times 7 by subtracting 2 tens and giving back 2 times 3. That's 70 minus 2 times 10 is 50. Add 2 times 3 to get 6 and the answer is 56.

Now do the calculation with the circles and you will see again it is the same calculation.

Does it matter if you or your children don't understand the explanation? Not in the least. Keep using the methods, learn your tables easily as you go, and you will suddenly find that the explanation does make sense. And, even if it doesn't make sense, your success won't be hindered in any way.

Conclusion

It is not the brain you are born with that is important, it is how you learn to use it. No-one is a born mathematician or a born problem-solver. Maths skills are developed. Problem-solving skills are developed. Intelligence is developed.

People who are good at mathematics are not necessarily more intelligent than everybody else, they just have better strategies. You have learnt some strategies in this book that can help you and your child become better mathematicians. Now it is up to you. Play with the strategies you have learnt. This is a fun way to develop your maths skills and to become a mathematician.

Learn your tables with your children by calculating the answer and doing it so quickly everyone will think you have them memorised. In fact, you and your children will probably have memorised up to the 20 or 30 times tables in a couple of months. How many others at their school can do that? This will also help your children because they will want to live up to the reputation they develop for themselves.

You will also be opening up new opportunities for your children. When jobs are hard to get, outstanding maths skills and a great deal of intelligence will give your children a head start in getting the job they want.

Factors that determine intelligence

Some people seem to have been born with a high level of intelligence and some of us seem to have less than our fair share. Many people look at others and what they can do, and think to themselves, 'I could never do that. They are a lot smarter than I am'. However, the fact is, intelligence is developed. We are not all born equal, but we could probably all do much better than we are currently doing.

I have worked with many high and low achievers. All have been told that they were stupid. The low achievers have it driven home time and time again. The high achievers are given much encouragement, so if they are told they are stupid, they don't accept it. They accept that they did something stupid, but not that they are stupid people. The low achievers believe it. After all, they have plenty of evidence to prove the statement.

We all do stupid things, but that doesn't make us stupid people. Einstein was told he was stupid. Einstein did stupid things and said some 'dumb' things, but that didn't make him a stupid person. He was an intelligent person who made some mistakes, as we all do.

How do we convince a low achiever he is not stupid, but highly intelligent? I use my maths and learning strategies to illustrate forcefully that it's not the brain you are born with that counts, but how you use it.

We are all born with the potential to be highly intelligent. Thinking skills are learnt. Some people learn very effective thinking and learning skills. Others learn clumsy and poor thinking skills. Few of us realise our true potential.

Thinking skills can affect our:

- learning ability
- schooling

- leisure time (hobbies, sports and so on)
- careers
- friendships or marriage
- raising of children.

We owe it to ourselves and our children to learn how to think effectively.

A number of factors determine intelligence. These include self-esteem, health and past performance. Helping your child work through these factors will help to improve her intelligence.

Self-esteem

If you believe you can solve a problem you are likely to work on it until you find a solution. However, if you believe you are stupid or lack intelligence, you will live out your belief. If you believe you are not smart enough, you won't persevere. It's not that you are lazy, or don't want to solve the problem, you just don't believe you will succeed. Telling a child, 'Try harder, try to do it', will just frustrate him. The child wants to try harder, but doesn't know how. What he needs is a strategy. What I do is teach children *how* to solve problems or do the calculations.

Health

If you have a headache, or just don't feel well, you don't perform so well mentally. If you are tired, have an ache, feel weary, or if you have an overriding concern or fear, you won't be able to concentrate, think clearly or solve problems effectively. But children are sometimes judged, by people who should know better, as being unintelligent, stupid or learning-disabled. Often a child is unable to properly judge the reasons for her lack of performance and accepts the opinion of her teacher, who says she lacks intelligence.

Past performance

If a child tries to solve a problem and can't solve it, he has started a pattern. It may simply be because the child didn't understand the teacher's explanation of what was required. Next time the child is given a similar problem, he says, 'I can't do this. This is hard. I failed last time because I can't do these problems. I'm not smart enough. I'm not as smart as the other kids who can do them'.

If a child doesn't understand the explanation given by the teacher, this is often regarded as evidence of low intelligence. It could, however, be due to the teacher's choice of words. A word may have one meaning for you and quite a different meaning for me. A teacher should explain a principle or problem from several points of view to help ensure all students understand what has been explained.

Teach a child how to solve puzzles, do multiplication, add or subtract, and you will start a pattern of success. Then the child will say, 'Here is another one of those problems. I can do these. I'm good at solving these. I did the last one. I can do this one'. We are not born with a high or low intelligence—intelligence can be developed.

Just a beginning

I hope the methods I have introduced prove useful for you and your children, and I wish you all the best for your future mathematical endeavours.

Finally, this book is only an introduction to my methods of teaching basic mathematics. After radio interviews, during which I give an example of how these methods work, people often telephone the radio station to say that the methods don't work for every case. I answer that they do, I just haven't taught the whole strategy in five minutes. Nor have I taught the whole strategy in this book as it is simply an introduction to maths.

If you are interested in learning more and building on what you have learnt in this book, you may want to buy the follow-up books or teachers' manual. If you would like to contact me, or find out about other learning materials, classroom materials and audio programs, you can write to me at 'Bill Handley, PO Box 545, Lilydale, VIC, 3140, Australia', via email at bhandley@speedmathematics .com, or visit my website www.speedmathematics.com. More practice sheets are also available online.

Key points

- Intelligence can be improved and developed.

- High achievers receive positive feedback, which helps to improve their performance.

- Low achievers receive negative feedback, which hinders their development.

- Self-esteem, health and past performance strongly influence a child's development.

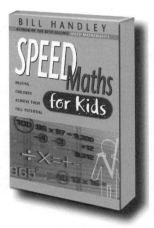